STUDENT UNIT GUIDE

UNIT

AC 4

C

Kinetics

Organic

Margaret Cross

I would like to acknowledge the original work by Neil Goldie, who wrote the AQA A2 Chemistry Unit 4 Guide for the previous specification. I have drawn extensively from his publication and I am grateful to Alison Goldie for giving me permission to adapt Neil's work for use in this guide to the new A2 specification.

Philip Allan Updates, an imprint of Hodder Education, an Hachette UK company, Market Place, Deddington, Oxfordshire OX15 0SE

Orders
Bookpoint Ltd, 130 Milton Park, Abingdon, Oxfordshire OX14 4SB
tel: 01235 827720
fax: 01235 400454
e-mail: uk.orders@bookpoint.co.uk
Lines are open 9.00 a.m.–5.00 p.m., Monday to Saturday, with a 24-hour message answering service. You can also order through the Philip Allan Updates website: www.philipallan.co.uk

ISBN 978-1-4441-0774-6

First printed 2010
Impression number 5 4 3 2 1
Year 2014 2013 2012 2011 2010

Printed by MPG Books, Bodmin

Contents

Introduction

■ ■ ■

Content Guidance

■ ■ ■

Questions & Answers

Introduction

About this guide

This guide is for students following the AQA Chemistry A2 specification. It deals with **Unit 4, Kinetics, Equilibria and Organic Chemistry**. This unit covers 20% of the total A2 marks. The examination lasts 1 hour 45 minutes and consists of six to eight short-answer structured questions, of varying length, and two or three longer, structured questions, which require answers written in continuous prose. Some of the questions will have synoptic elements. All questions are compulsory. There are 100 marks available on this paper.

Questions in Unit 4 assume knowledge and understanding of the concepts of physical and organic chemistry covered at AS. In addition, there are many principles and concepts covered in Unit 4 that must be understood so that you can answer questions set in an unfamiliar situation.

The key to success

It is essential that you can recall the basic facts and definitions, but a deeper understanding of the subject is essential if you want to achieve the top grades. The key to success in chemistry is to understand the fundamental concepts and be able to apply them to new and unfamiliar situations. Good examination technique is also an important factor, enabling you to work more effectively in the exam and gain access to the marks needed for a top grade.

This guide allows you to look again at the content of the specification, to test yourself at the end of each section and to assess your own work. It is essential that you read through the examiner's comments because they will help you improve your exam technique. Once you have completely worked through this guide, you will be very aware of your weak areas and it is these areas that you need to address. Make a list of these weak areas and discuss any problems with other students in your class and with your teacher.

Using the guide

This guide has three sections:

- **Introduction** — this provides guidance on study and revision, together with advice on approaches and techniques to ensure you answer the examination questions in the best way that you can.
- **Content Guidance** — this section is not intended to be a textbook. It offers guidelines on the main features of the content of Unit 4, together with particular advice on making study more productive.
- **Questions and Answers** — this shows you the sort of questions you can expect in the unit test. Grade-A answers are provided; these are followed by examiner's

comments. Careful consideration of these will improve your answers and, much more importantly, will improve your understanding of the chemistry involved.

Revise a topic using the Content Guidance section as a guide. If there is something you do not understand, you should also refer to your own class notes and textbooks.

It is essential that you write down *specific* questions and discuss them with your teacher. For instance, 'Please could you explain equilibria again' is not a good use of your time if you understand most of the features of equilibria. 'Please could you explain K_c and calculations involving K_c' is more specific and shows that you have worked hard to identify weak areas.

Once you have revised a particular topic thoroughly, you should attempt the relevant questions in the Question and Answer section, *without* looking at the grade-A answer.

Compare your answer with the grade-A answer and estimate your own performance. A rough guide to use is 80% = grade A, 70% = grade B, 60% = grade C, etc. However, these grade boundaries are adjusted, depending on the individual paper and the performance of the candidates.

Read through the examiner's comments to find out if you have made any of the common mistakes and to see how you could improve your technique. The comments also give some alternative answers.

Make a note of *specific questions* that caused you problems and discuss them with other students and your teacher.

Revision schedule

- Plan your revision schedule carefully.
- It is essential that you revise regularly.
- Leave yourself enough time to cover all the material. You need to go through each topic once as a basic minimum, and then go through the weak areas again.
- In the weeks leading up to the exam, it is the weak areas that you should be revising, not every topic.
- In each revision session do not try to achieve too much. Revise one topic per session, e.g. bond enthalpies. Here is one way to structure your session:
  - revise from the Content Guidance section (and your own notes)
  - make a brief written summary (no more than an A4 sheet of paper)
  - attempt the questions
  - mark your answer
  - read the examiner's comments
- If you score a grade A, tick the relevant section of the specification. You must have a break before you start to revise the next topic.
- If there are weak areas and questions that you clearly do not understand, then write down specific questions ready for discussion with your teacher.

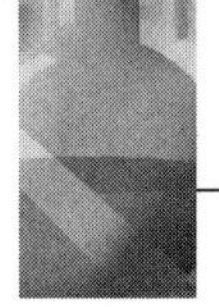

- Finally, make sure that you attempt some past-paper questions from the exam board and study the mark schemes carefully. Questions may be repeated or similar questions set.

Unit Test 4

If you have revised thoroughly, completed all the questions in this guide and discussed problems with other students and your teacher, you should enjoy the exam. If you have completed some AQA past papers, then the style of the paper will be familiar and you will recognise some questions in the exam because they will be similar to previous questions.

Do not begin to write as soon as you open the paper — quickly scan the questions first.

It is *not* essential that you answer the questions in order. If the first question is difficult, leave it to the end. It *is* essential that you answer *all* the questions.

You will have enough time to answer all the questions, provided you keep your answers concise and you do not include irrelevant information. It is easy to waste time writing out a section of your notes that is irrelevant to the question. Do not repeat the question when starting your answer. The key to exam success is achieving the maximum number of marks in the minimum number of words.

The mark allocation at the end of each question should be used to estimate the amount of detail needed in your answer. If there is 1 mark available, the examiner will be looking for a key word or phrase and certainly no more than one sentence. If there are 4 marks available, you should include four key points, which usually means writing four short sentences.

You will be assessed on your quality of written communication (QWC) in Section B of the exam paper. This will assess your ability to:

- support claims with an appropriate range of evidence
- show good use of information or ideas about chemistry
- give well-structured arguments with minimal repetition or irrelevant points
- express ideas accurately and clearly with only minor errors of grammar, punctuation and spelling

There are no discrete marks for the assessment of QWC but the examiner will consider these points before awarding the marks for your answers.

No marks are available for producing neat answers, but it certainly helps the examiners when they are marking your work. Untidy diagrams may become inaccurate and this definitely loses marks.

Content Guidance

This section covers the content of **Unit 4: Kinetics, Equilibria and Organic Chemistry**. In this Content Guidance section, the specification has been converted into user-friendly language and is in a format that is easy to remember. All the key facts, definitions and basic principles are covered but the section is not intended to be a textbook and you should consult a standard text and your notes for further detail. In order to achieve a top grade, it is essential that you understand fully the basic concepts and that you can apply them to unfamiliar situations.

The content of this unit falls into 11 sections, which are summarised below.

Kinetics: simple rate equations; determination of rate equation.

Equilibria: equilibrium constant K_c for homogeneous systems; qualitative effects of changes of temperature and concentration.

Acids and bases: Brønsted–Lowry acid–base equilibria in aqueous solution; definition and determination of pH; the ionic product of water, K_w; weak acids and bases, K_a for weak acids; pH curves, titrations and indicators; buffer action.

Nomenclature and isomerism in organic chemistry: naming organic compounds; isomerism.

Compounds containing the carbonyl group: aldehydes and ketones; carboxylic acids and esters; acylation.

Aromatic chemistry: bonding; delocalisation stability; electrophilic substitution; nitration; Friedel–Crafts acylation reactions.

Amines: base properties (Brønsted–Lowry); nucleophilic properties; preparation.

Amino acids: acid and base properties; proteins.

Polymers: addition polymers; condensation polymers; biodegradability and disposal of polymers.

Organic synthesis and analysis: applications.

Structure determination: data sources; mass spectrometry; infrared spectroscopy; nuclear magnetic resonance spectroscopy; chromatography.

Kinetics

This topic was covered in AS Unit 2, which concentrated on the factors that are essential for a successful reaction: collision, activation energy and orientation. Unit 2 also looked at the *qualitative* effect on reaction rate of changing conditions (concentration, temperature and the addition of a catalyst). All this knowledge is assumed here.

The kinetics section in Unit 4 is more concerned with the *quantitative* relationship between the rate of the reaction and the concentration of the reactants, which is shown by the rate equation.

Simple rate equations

The rate equation shows the relationship between the rate of reaction and the concentration of each reactant.

$r = k[A]^m[B]^n$

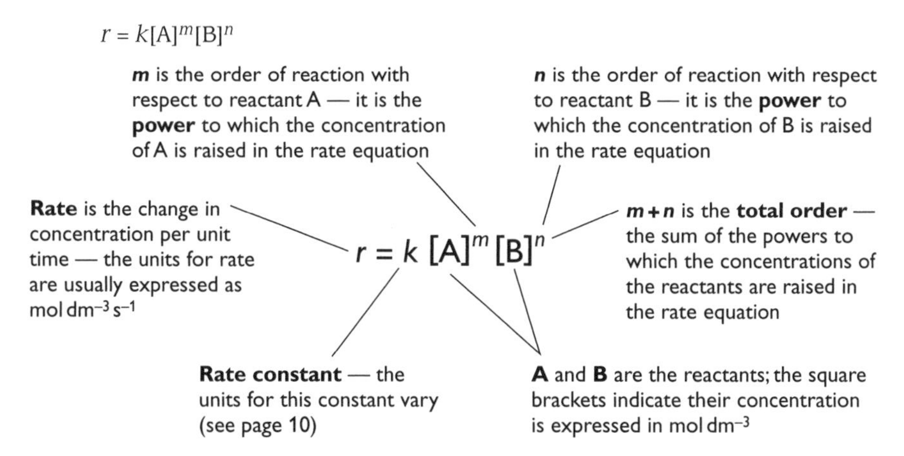

Deriving a rate equation

The effect of concentration of a reactant on the initial rate of reaction can only be derived by experiment, it cannot be determined from the stoichiometry of the equation. Consider the following example:

$$A + 2B \longrightarrow C + D$$

reactants products

The initial rate of reaction was measured at constant temperature. The concentrations of A and B were varied and the following results obtained.

Experiment	Initial concentration of A/mol dm^{-3}	Initial concentration of B/mol dm^{-3}	Initial rate of reaction/ mol dm^{-3} s^{-1}
1	1.0	1.0	2×10^{-4}
2	2.0	1.0	8×10^{-4}
3	1.0	2.0	4×10^{-4}
4	2.0	2.0	Unknown

In experiments 1 and 2, the concentration of reactant A is doubled and the concentration of B is constant. The rate of the reaction increases by a factor of 4 (from 2×10^{-4} mol dm^{-3} s^{-1} to 8×10^{-4} mol dm^{-3} s^{-1}). The rate is proportional to $[A]^2$ and the **order is 2** with respect to reactant A.

In experiments 1 and 3, the concentration of reactant B is doubled and the concentration of A is constant. The rate of the reaction increases by a factor of 2 (from 2×10^{-4} mol dm^{-3} s^{-1} to 4×10^{-4} mol dm^{-3} s^{-1}). The rate is proportional to $[B]^1$ and the **order is 1** with respect to reactant B.

The **total order** of the reaction is 2 + 1 = 3. This means that the rate is proportional to [concentration]^3. When comparing experiments 1 and 4, the concentration of both reactants has doubled, so the rate would be expected to go up by a factor of 2^3, i.e. 8, so the unknown rate will be 16×10^{-4} or 1.6×10^{-3} mol dm^{-3} s^{-1}.

The rate equation for this reaction is:

$$r = k[A]^2[B]^1$$

Calculating the rate constant, *k*

Rearranging the rate equation gives:

$$k = \frac{r}{[A]^2[B]^1}$$

The rate constant can then be calculated by taking the data from any experiment. For example, using the data from experiment 1:

$$k = \frac{2 \times 10^{-4}}{1.0^2 \times 1.0} = 2 \times 10^{-4} \text{ mol}^{-2}\text{ dm}^6\text{ s}^{-1}$$

In experiment 4, the rate constant can be used to predict the unknown rate of reaction.

$$r = k[A]^2[B]^1 = 2 \times 10^{-4} \times 2.0^2 \times 2.0 = 2 \times 10^{-4} \times 8 = 1.6 \times 10^{-3} \text{ mol dm}^{-3}\text{ s}^{-1}$$

Units of the rate constant, *k*

The units of the rate constant vary according to the rate equation. Here a reaction of order 3 is used to show how to work out the units of *k*.

- A typical rate equation is $r = k[A]^2[B]$
- This can be rearranged as: $k = \dfrac{r}{[A]^2[B]}$

- Write out the units and cancel where possible: $\frac{\cancel{\text{mol dm}^{-3}}\,\text{s}^{-1}}{\cancel{\text{mol dm}^{-3}}\,\text{mol dm}^{-3}\,\text{mol dm}^{-3}}$
- Units of k: $\text{mol}^{-2}\,\text{dm}^{6}\,\text{s}^{-1}$

Remember:

- the units for rate are usually $\text{mol dm}^{-3}\,\text{s}^{-1}$
- the units for concentration are mol dm^{-3}
- when transferring mol^{2} from the bottom line to the top line it becomes mol^{-2}
- transferring dm^{-6} from the bottom line to the top line becomes dm^{6}

The effect of temperature on the rate constant, *k*

An increase in temperature always leads to an increase in the rate of the reaction. An increase in temperature causes an increase in the average kinetic energy of the particles, so they move faster and this leads to an increased collision frequency. However, a more important factor is that the increased energy means more particles now exceed the energy of activation (the minimum energy for a reaction to occur), so there are many more successful collisions. This leads to a dramatic increase in the rate of the reaction.

In the rate equation, $r = k[\text{A}]^{m}[\text{B}]^{n}$, the only temperature-dependent feature is the rate constant. The rate constant increases exponentially with an increase in temperature.

Reaction mechanisms and the rate-determining step

Many reactions occur in a series of steps and only those reactants that take part in the slowest step (and any steps before the slowest step) will appear in the rate equation. It is the rate of the *slowest* step that is measured during a kinetic study; this step is the **limiting factor** for the rate of reaction as a whole and is known as the **rate-determining step**.

The mechanism of a reaction is the sequence of steps by which the reaction is believed to occur. A proposed mechanism must be consistent with the observed kinetic data and therefore the kinetic data for a reaction allow chemists to speculate about the mechanism by which the reaction takes place.

Example

The reaction of bromine with propanone in the presence of hydroxide ions can be represented by the overall equation:

$$CH_3COCH_3(aq) + Br_2(aq) + OH^-(aq) \longrightarrow CH_3COCH_2Br(aq) + H_2O(l) + Br^-(aq)$$

The rate equation deduced by experiment is:

$$\text{rate} = k[CH_3COCH_3(aq)][OH^-(aq)]$$

The rate-determining step does not involve bromine since changing the concentration of bromine does not affect the rate of the reaction. The proposed mechanism for the reaction involves two steps.

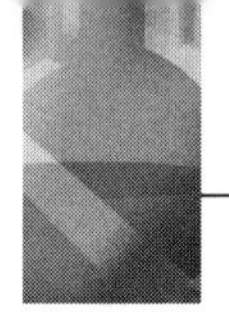

Step 1: $CH_3COCH_3(aq) + OH^-(aq) \longrightarrow CH_3COCH_2^-(aq) + H_2O(l)$
Step 2: $CH_3COCH_2^-(aq) + Br_2(aq) \longrightarrow CH_3COCH_2Br(aq) + Br^-(aq)$

Step 1 is the slow rate-determining step. Step 2 is very fast.

Note: you are not expected to recall examples of the rate-determining step for reactions; sufficient information will always be supplied for you to apply the principles to unfamiliar reactions.

Equilibria

You must be able to apply the concepts of equilibrium reactions met in Unit 2 and extend these to include quantitative aspects of equilibria.

Equilibrium constant, K_c, for homogeneous systems

Consider the following equilibrium:

$$aA + bB \rightleftharpoons cC + dD$$

The equilibrium constant K_c is defined by the expression:

$$K_c = \frac{[C]^c[D]^d}{[A]^a[B]^b}$$

- 'Homogeneous' means all the species are in the same phase.
- The equilibrium constant K_c is calculated from the concentrations in mol dm^{-3} at constant temperature.

Expressions for K_c for some common equilibrium reactions

Reaction	K_c	Units
$PCl_5 \rightleftharpoons PCl_3 + Cl_2$	$K_c = \frac{[PCl_3][Cl_2]}{[PCl_5]}$	mol dm^{-3}
$H_2 + I_2 \rightleftharpoons 2HI$	$K_c = \frac{[HI]^2}{[H_2][I_2]}$	No units
$N_2 + 3H_2 \rightleftharpoons 2NH_3$	$K_c = \frac{[NH_3]^2}{[N_2][H_2]^3}$	mol^{-2} dm^6
$2SO_2 + O_2 \rightleftharpoons 2SO_3$	$K_c = \frac{[SO_3]^2}{[SO_2]^2[O_2]}$	mol^{-1} dm^3

- The larger the value of K_c the more products will be present in the equilibrium mixture.
- The units of K_c vary according to the number of reactants and products. If the numbers of moles of reactants and products are the same, then K_c will have no units.
- The value of K_c is dependent on temperature. In an *endothermic* reaction, an increase in temperature will cause an *increase* in the value of K_c. In an *exothermic* reaction, an increase in temperature will cause a *decrease* in the value of K_c.
- A change in concentration will alter the equilibrium position, but not the value of the equilibrium constant.
- A catalyst has no effect on the equilibrium position or the equilibrium constant.

Calculations involving K_c

Example: the decomposition of phosphorus(V) chloride

1.33 mol of phosphorus(V) chloride vapour was heated to 500 K in a vessel of volume 15 dm^3. The equilibrium mixture contained 0.80 mol of chlorine. Calculate the value of K_c for this decomposition into PCl_3 and Cl_2.

	PCl_5	$\rightleftharpoons$	PCl_3	+	Cl_2
Initial concentration	$\frac{1.33}{V}$		0		0
Equilibrium concentrations	$\frac{1.33 - x}{15}$		$\frac{x}{15}$		$\frac{x}{15}$

The value of *x* is 0.80 because this is the number of moles of chlorine in the equilibrium mixture (this is given in the question). *V* is 15 dm^3 and is the total volume of the equilibrium mixture. In this calculation, where the number of reactant particles does not equal the number of product particles, *V* does not cancel.

$$[PCl_5] = \frac{1.33 - 0.80}{15} = 0.035$$

$$[PCl_3] = \frac{0.80}{15} = 0.053$$

$$[Cl_2] = \frac{0.80}{15} = 0.053$$

$$K_c = \frac{[PCl_3][Cl_2]}{[PCl_5]} = \frac{0.053 \times 0.053}{0.035} = 0.080\ \text{mol dm}^{-3}$$

Changing the conditions of an equilibrium reaction

The qualitative effect on the position of equilibrium of changing the reaction conditions can be predicted by using **Le Chatelier's principle** and has been met in Unit 2.

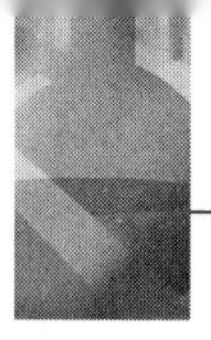

The effect of a change in concentration

The Haber process can be used as an example:

$N_2(g) + 3H_2(g) \rightleftharpoons 2NH_3(g)$

- If the concentration of either reactant N_2 or H_2 is increased, then the position of the equilibrium is displaced to the right (so as to reduce the concentration of N_2 or H_2). This means that more NH_3 is obtained.
- If the product NH_3 is removed, the position of the equilibrium is also displaced to the right to replace the NH_3.
- The equilibrium constant K_c is unchanged, since there is no change in temperature.

The effect of a change in temperature

- If the temperature is changed, the position of the equilibrium shifts to oppose the change.
- If the reaction is exothermic in the forward direction, an increase in temperature would shift the position of the equilibrium to the left and there would be fewer products in the equilibrium mixture. The converse is true if the temperature is decreased.
- If the reaction is endothermic in the forward direction, an increase in temperature would shift the position of the equilibrium to the right and there would be more products in the equilibrium mixture. The converse is true if the temperature is decreased.
- If the temperature is changed, the value of K_c will change. The direction of the change depends on whether the (forward) reaction is exothermic or endothermic.

Example 1: decomposition of dinitrogen tetraoxide

$N_2O_4 \rightleftharpoons 2NO_2 \qquad \Delta H = +58\,\text{kJ mol}^{-1}$

This reaction is endothermic in the forward direction.

- If the temperature is increased, the equilibrium responds to try to reduce the temperature.
- The equilibrium position shifts to the right (in the endothermic direction) and more NO_2 is produced.
- The value of K_c increases as the temperature increases.

Example 2: the manufacture of methanol

$CO(g) + 2H_2(g) \rightleftharpoons CH_3OH(g) \qquad \Delta H = -91\,\text{kJ mol}^{-1}$

This reaction is exothermic in the forward direction.

- If the temperature is increased, the equilibrium responds to try to reduce the temperature.
- The equilibrium position shifts to the left (in the endothermic direction) and less CH_3OH is produced.
- The value of K_c decreases as the temperature increases.

The effect of the addition of a catalyst

- The addition of a catalyst to a mixture at equilibrium has no effect on the composition of the equilibrium mixture.
- A catalyst speeds up the rate of the forward and backward reactions equally.

- A catalyst has no effect on the equilibrium *position* but the *rate* of attainment of equilibrium is increased.
- A catalyst has no effect on the value of the equilibrium constant, K_c.

Acids and bases

The Brønsted–Lowry theory of acids and bases

- An acid is a proton (H^+) donor.
- A base is a proton (H^+) acceptor.
- Acid–base equilibria involve the transfer of protons.

Example 1: an acid with water

The acid HA reacts with water to produce the following equilibrium:

$$\underset{\text{acid}}{HA} + \underset{\text{base}}{H_2O} \rightleftharpoons \underset{\text{base}}{A^-} + \underset{\text{acid}}{H_3O^+}$$

Example 2: a base with water

The base A^- reacts with water to produce the following equilibrium:

$$\underset{\text{base}}{A^-} + \underset{\text{acid}}{H_2O} \rightleftharpoons \underset{\text{acid}}{HA} + \underset{\text{base}}{OH^-}$$

You might be asked to complete equations and label the species as acids or bases. Common examples include the following:

$$\underset{\text{acid}}{HCl} + \underset{\text{base}}{H_2O} \rightleftharpoons \underset{\text{acid}}{H_3O^+} + \underset{\text{base}}{Cl^-}$$

$$\underset{\text{base}}{NH_3} + \underset{\text{acid}}{H_2O} \rightleftharpoons \underset{\text{acid}}{NH_4^+} + \underset{\text{base}}{OH^-}$$

The definition and calculation of pH

The acidity of an aqueous solution depends on the number of $H^+(aq)$ ions in solution.

- pH is defined by the equation:

  $pH = -\log_{10}[H^+]$

  where $[H^+]$ is in $mol\,dm^{-3}$.
- Acid strength depends on the concentration of H^+ ions in $mol\,dm^{-3}$; the greater the concentration of H^+ ions, the lower the pH and the greater the acid strength.
- When calculating the pH of a strong acid (e.g. HCl, HNO_3, H_2SO_4) or a strong base (e.g. NaOH, KOH) complete dissociation is always assumed.

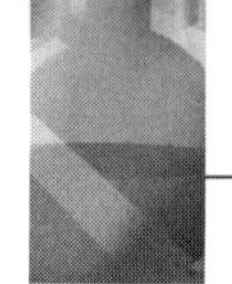

A similar relationship exists for a strong base, the strength of the base being dependent on the concentration of OH^- ions in mol dm^{-3} in solution.

Examples

(1) Calculate the pH of an aqueous solution of 0.100 mol dm^{-3} HCl. Assuming complete dissociation, 0.100 mol dm^{-3} HCl gives 0.100 mol dm^{-3} H^+ ions.

$pH = -\log_{10}[H^+] = -\log_{10}0.100 = 1.00$

(2) Calculate the concentration of H^+ ions given that the pH of an aqueous solution of HCl is 2.50.

$pH = -\log_{10}[H^+]$, so $[H^+] = 10^{-pH}$

$[H^+] = 10^{-2.5} = 3.16 \times 10^{-3}$ mol dm^{-3}

The ionic product of water, K_w

Water can act as both an acid (a proton donor) and a base (a proton acceptor). As a result, both H_3O^+ ions and OH^- ions can exist simultaneously in water.

$H_2O + H_2O \rightleftharpoons H_3O^+ + OH^-$

This equilibrium can be represented more simply as:

$H_2O(l) \rightleftharpoons H^+(aq) + OH^-(aq)$

The equilibrium constant, K_c, for this reaction is:

$$K_c = \frac{[H^+][OH^-]}{[H_2O]}$$

The amount of water that dissociates is incredibly small and the equilibrium lies well over to the left-hand side. The concentration of water can be treated as a constant and incorporated into the value of K_c. This produces a new constant, called the ionic product of water, K_w.

$K_w = [H^+][OH^-] = 1 \times 10^{-14}$ mol^2 dm^{-6} at 298 K

In neutral solution $[H^+] = [OH^-]$, so at 298 K $[H^+] = [OH^-] = 1 \times 10^{-7}$

$\therefore pH = -\log_{10}[H^+] = -\log_{10}1 \times 10^{-7} = 7.00$ at 298 K

The value of K_w increases with an increase in temperature. This is because the dissociation of water is an endothermic reaction (+57 kJ mol^{-1}).

The value of K_w can be used to calculate the pH of a solution of a strong base from its concentration.

Examples

(1) Calculate the pH of a 0.100 mol dm^{-3} solution of sodium hydroxide.

$K_w = [H^+][OH^-] = 1 \times 10^{-14}$ mol^2 dm^{-6}

Assuming complete dissociation, $[OH^-] = 0.100$ mol dm^{-3}

$$[H^+] = \frac{K_w}{[OH^-]} = \frac{1 \times 10^{-14}}{0.100} = 1 \times 10^{-13} \text{ mol dm}^{-3}$$

$pH = -\log_{10}[H^+] = -\log_{10}(1 \times 10^{-13}) = 13.00$

(2) An aqueous solution of potassium hydroxide has a pH of 12.90. Calculate the concentration in $mol\,dm^{-3}$ of the potassium hydroxide solution.

$pH = -\log_{10}[H^+]$, so $[H^+] = 10^{-pH}$

$[H^+] = 10^{-12.9} = 1.26 \times 10^{-13}\,mol\,dm^{-3}$

$$[OH^-] = \frac{K_w}{[H^+]} = \frac{1 \times 10^{-14}}{1.26 \times 10^{-13}} = 0.0790\,mol\,dm^{-3}$$

Since potassium hydroxide is a strong base, its concentration will also be $0.0790\,mol\,dm^{-3}$.

Weak acids and bases

A weak acid only partially dissociates in aqueous solution. An example of a weak acid is ethanoic acid, CH_3COOH.

$$CH_3COOH(aq) \rightleftharpoons CH_3COO^-(aq) + H^+(aq)$$

The amount of dissociation is indicated by the acid dissociation constant, K_a.

$$K_a = \frac{[CH_3COO^-][H^+]}{[CH_3COOH]} = 1.70 \times 10^{-5}\,mol\,dm^{-3}$$

The larger the K_a value of an acid, the greater the dissociation of the acid and the stronger the acid.

- HF, with a K_a value of $5.60 \times 10^{-4}\,mol\,dm^{-3}$, is a stronger acid than ethanoic acid.
- HCN, with a K_a value of $4.90 \times 10^{-10}\,mol\,dm^{-3}$, is a weaker acid than ethanoic acid.

A weak base only partially dissociates in aqueous solution. An example of a weak base is ammonia:

$$NH_3(aq) + H_2O(l) \rightleftharpoons NH_4^+(aq) + OH^-(aq)$$

The definition of pK_a

Acid strength can also be defined by the term pK_a, where p$K_a = -\log_{10}K_a$.

Consider the following K_a and pK_a values:

acid A	$K_a = 1.00 \times 10^{-5}\,mol\,dm^{-3}$	p$K_a = 5.00$
acid B	$K_a = 1.00 \times 10^{-3}\,mol\,dm^{-3}$	p$K_a = 3.00$

The stronger acid will dissociate more, leading to a larger K_a value and a smaller pK_a value. Acid B is stronger than acid A.

Calculating the pH of a weak acid

Using the equation for the dissociation of the weak acid, CH_3COOH, it can be seen that the number of moles of H^+ ions produced is always the same as the number of moles of CH_3COO^- ions produced. It is assumed that $[H^+]$ is solely due to the dissociation of the acid, so the top line of the K_a expression becomes $[H^+]^2$. For weak acids it can also be assumed that the amount of dissociation is so small that the concentration of the CH_3COOH at equilibrium is the same as the original concentration of CH_3COOH.

Example

Calculate the pH of a 0.100 mol dm^{-3} solution of ethanoic acid (K_a = 1.70 × 10^{-5} mol dm^{-3}).

$$1.70 \times 10^{-5} = \frac{[CH_3COO^-][H^+]}{[CH_3COOH]} = \frac{[H^+]^2}{[CH_3COOH]_{original}} = \frac{[H^+]^2}{0.100}$$

$$[H^+]^2 = 0.100 \times 1.70 \times 10^{-5} = 1.70 \times 10^{-6}$$

$$\therefore \quad [H^+] = \sqrt{1.70 \times 10^{-6}} = 1.304 \times 10^{-3}$$

$$pH = -\log_{10}(1.304 \times 10^{-3}) = 2.88$$

Note: in the above example the pK_a value (= 4.75) could have been given, in which case there would have been an additional first step:

$$K_a = 10^{-4.75} = 1.70 \times 10^{-5}$$

pH curves, titrations and indicators

pH curves

A graph of the pH of a solution being titrated against the volume of solution added is known as a pH curve. The **equivalence point** (or stoichiometric point) occurs when stoichiometrically equivalent amounts of acid and base have been added together. The **end point** of the titration is the point at which the indicator changes colour. The equivalence point can be determined accurately with an indicator when the end point coincides with the equivalence point. An indicator is suitable if the rapid change of pH at equivalence (shown by the near-vertical portion on the pH curve) overlaps the range of activity of the indicator.

The following pH curves are for various combinations of 0.100 mol dm^{-3} solutions of acids and bases. The ranges of two common indicators, phenolphthalein (range = 8.3–10.0) and methyl orange (range = 3.2–4.4), are shown to demonstrate their suitability. Summaries are provided for each titration curve. You need to be able to sketch these curves. Always remember to label the axes. The symbols used on the pH curves are EP (equivalence point), PP (phenolphthalein) and MO (methyl orange).

Addition of a strong base to a strong acid

$$HCl(aq) + NaOH(aq) \longrightarrow NaCl(aq) + H_2O(l)$$

- The pH starts at 1 and rises slowly.
- There is a rapid change in pH just before the equivalence point.
- The pH is 7 at the equivalence point.
- The pH is 12.5 after adding 50 cm^3 of NaOH.
- Either methyl orange or phenolphthalein could be used to show the equivalence point.

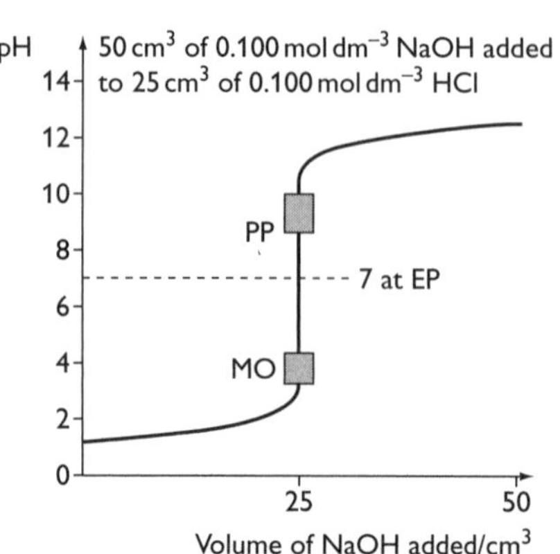

Addition of a strong base to a weak acid

$CH_3COOH(aq) + NaOH(aq) \longrightarrow CH_3COONa(aq) + H_2O(l)$

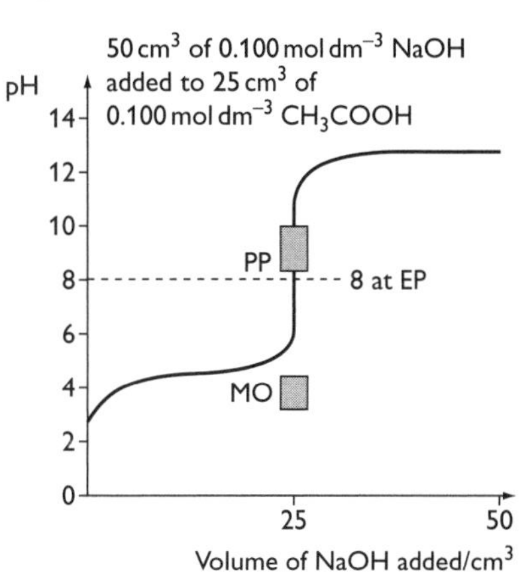

- The pH starts at approximately 3 and rises slowly.
- There is a rapid change in pH just before the equivalence point.
- The pH is approximately 8 at the equivalence point due to the presence of OH^- ions in solution. These OH^- ions come from the reaction:
  $CH_3COO^-(aq) + H_2O(l) \rightleftharpoons CH_3COOH(aq) + OH^-(aq)$
- The pH is 12.5 after adding 50 cm^3 of NaOH.
- Phenolphthalein can be used to indicate the equivalence point; methyl orange is not suitable.

Addition of a weak base to a strong acid

$HCl(aq) + NH_3(aq) \longrightarrow NH_4Cl(aq)$

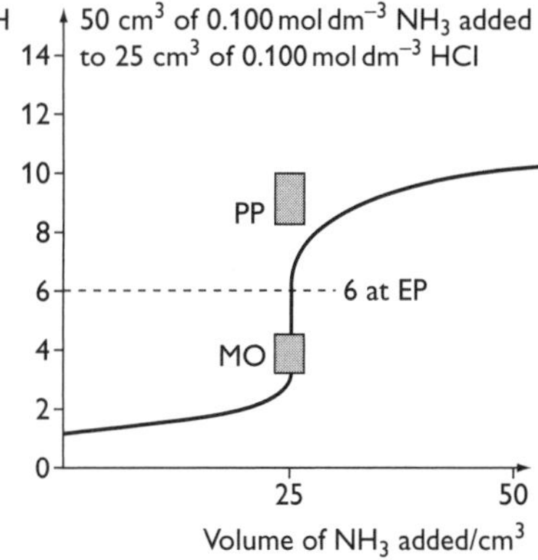

- The pH starts at 1 and rises slowly.
- There is a rapid change in pH just before the equivalence point.
- The pH is approximately 6 at the equivalence point due to the presence of H^+ ions in solution. These H^+ ions come from the reaction:
  $NH_4^+(aq) + H_2O(l) \rightleftharpoons NH_3(aq) + H_3O^+(aq)$
- The pH is approximately 11 after adding 50 cm^3 of NH_3.
- Methyl orange can be used to indicate the equivalence point; phenolphthalein is not suitable.

Addition of a weak base to a weak acid

$CH_3COOH(aq) + NH_3(aq) \longrightarrow CH_3COO^-(aq) + NH_4^+(aq)$

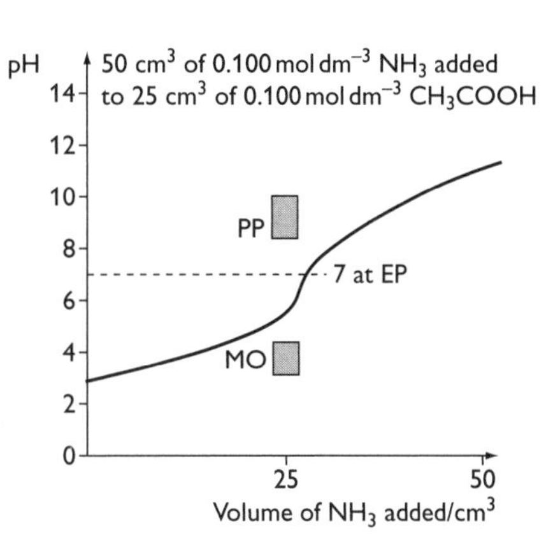

- The pH starts near 3 and rises gradually.
- The variation of pH with volume near the end point is too gradual for the detection of the equivalence point (at pH 7).
- The pH is around 11 after adding 50 cm^3 NH_3.
- Neither methyl orange nor phenolphthalein is suitable.

Note: in all these titration curves the alkali is added to the acid. Questions can be set relating to curves for titrations in which an acid is added to an alkali. In this case, the pH is above 7 at the start and below 7 at the end, i.e. the above curves are inverted.

Using a pH curve to determine pK_a and K_a for a weak acid

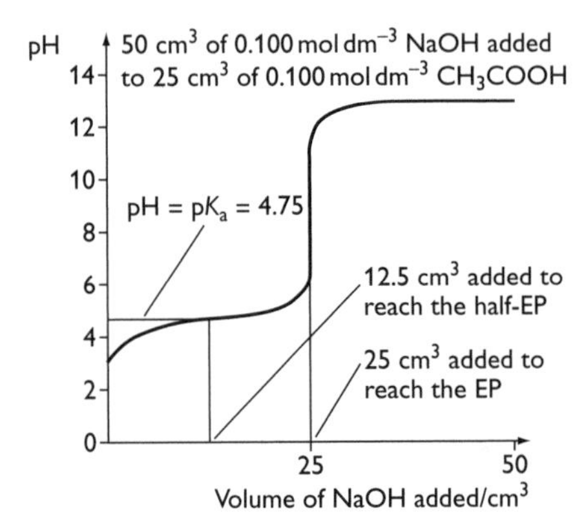

The pH curve for a weak acid and a strong base can be used to determine pK_a. If the volume needed for the equivalence point is 25.0 cm^3, then the half-equivalence point occurs at 12.5 cm^3. This is the point when only half the acid has been neutralised. At this point $[CH_3COOH] = [CH_3COO^-]$, so the expression for K_a can be simplified.

$$K_a = \frac{[CH_3COO^-][H^+]}{[CH_3COOH]}$$

$K_a = [H^+]$, so $\mathbf{p}\boldsymbol{K_a}\ \mathbf{= pH}$

Reading from the graph, the pH = 4.75, so pK_a = 4.75 and K_a is 1.78×10^{-5} mol dm^{-3}.

Indicators

An acid–base indicator is a water-soluble weak organic acid. The acid and base forms of the indicator, which can be represented by HIn and In^-, have different colours and the colour changes over a narrow pH range. The two forms exist in the following equilibrium:

HIn	⇌	In^-	+	H^+
Colour 1, at low pH, acidic conditions		Colour 2, at high pH, alkaline conditions		

At the equivalence point of an acid–base titration, the pH changes rapidly through several units of pH. The above equilibrium swings from almost all HIn to virtually all In^- (or vice versa, depending on which solution is in the burette) so the colour changes from colour 1 to colour 2. For example, for phenolphthalein this would be from colourless to pink. This colour change is used to indicate the end point of a titration. The colour change will occur over a different narrow pH range for different indicators and if it coincides with the rapid change of pH at the equivalence point, then the indicator is suitable for use in the titration.

An indicator is a weak acid and the pH range at which it changes colour is determined by its acid dissociation constant, K_a, which for indicators is usually represented by K_{in}.

$$K_{in} = \frac{[H^+][In^-]}{[HIn]}$$

At the end point, $[HIn] = [In^-]$, so $K_{in} = [H^+]$. This means that pK_{in} = pH at the end point. The pH range over which the colour change occurs is approximately $pK_{in} \pm 1$ unit.

Different indicators are used because they cover different ranges of pH. The pH ranges and colour changes for some common acid–base indicators are given in the table.

Indicator	pK_{in}	pH range	Colour in acid	Colour in alkali
Methyl orange	3.7	3.2–4.4	Red	Yellow
Methyl red	5.1	4.2–6.3	Red	Yellow
Bromothymol blue	7.0	6.0–7.6	Yellow	Blue
Phenolphthalein	9.3	8.3–10.0	Colourless	Pink

Calculations for acid–base titrations

Concentrations and volumes of reaction

13.30 cm^3 of a solution of NaOH was required to neutralise 20.0 cm^3 of a 0.0800 mol dm^{-3} solution of H_2SO_4. Calculate the concentration, in mol dm^{-3}, of the NaOH solution.

$$2NaOH + H_2SO_4 \longrightarrow Na_2SO_4 + 2H_2O$$

$$\text{moles of } H_2SO_4 = \frac{\text{conc.} \times \text{volume}}{1000} = \frac{0.0800 \times 20.0}{1000} = 1.60 \times 10^{-3}$$

From the equation, 2 moles NaOH react with 1 mole H_2SO_4.

$$\text{moles of NaOH in } 13.30\,cm^3 = 2 \times 1.60 \times 10^{-3} = 3.20 \times 10^{-3}$$

$$\text{moles of NaOH in } 1000\,cm^3 = \frac{3.20 \times 10^{-3} \times 1000}{13.3} = 0.240\text{ mol NaOH}$$

$$\text{concentration of NaOH} = 0.240\text{ mol dm}^{-3}$$

Calculating the pH during a strong acid–strong base titration

These are the stages to be followed in calculations of this type.

- Calculate the number of moles of acid using $\frac{\text{conc.} \times \text{volume}}{1000}$
- Calculate the number of moles of base using $\frac{\text{conc.} \times \text{volume}}{1000}$
- Calculate the number of moles of excess acid (or excess base).
- Find the total volume of solution produced by the mixture (acid + base).
- Determine the concentration of excess acid (or base) in mol dm^{-3}.
- Determine the pH using $pH = -\log_{10}[H^+]$.
- Use $K_w = [H^+][OH^-]$ if calculating the pH of a base.

Calculating the pH during a weak acid–strong base titration

The method depends on how far the titration has progressed.

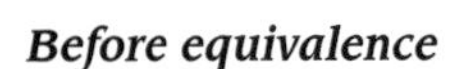

Before equivalence

The relative proportions of the weak acid and its anion have to be determined and the values used in the expression for K_a.

Calculate the pH in a titration when $10.00\,cm^3$ of $0.200\,mol\,dm^{-3}$ solution of NaOH has been added to $25.0\,cm^3$ of a $0.250\,mol\,dm^{-3}$ solution of ethanoic acid ($K_a = 1.76 \times 10^{-5}\,mol\,dm^{-3}$).

$$CH_3COOH + NaOH \longrightarrow CH_3COONa + H_2O$$

or more simply:

$$CH_3COOH + OH^- \rightleftharpoons CH_3COO^- + H_2O$$

$$\text{original moles of } CH_3COOH = \frac{\text{conc.} \times \text{volume}}{1000} = \frac{0.250 \times 25.0}{1000} = 6.25 \times 10^{-3}$$

$$\text{moles of NaOH added} = \frac{\text{conc.} \times \text{volume}}{1000} = \frac{0.200 \times 10.0}{1000} = 2.00 \times 10^{-3}$$

moles of CH_3COO^- ions formed $= 2.00 \times 10^{-3}$

moles of CH_3COOH remaining $= (6.25 \times 10^{-3}) - (2.00 \times 10^{-3}) = 4.25 \times 10^{-3}$

Since the ethanoic acid and the ethanoate ions exist together in the same overall volume, the ratio of the concentrations is equal to the ratio of the number of moles.

$$K_a = \frac{[CH_3COO^-][H^+]}{[CH_3COOH]}$$

$$\text{so } [H^+] = K_a \times \frac{[CH_3COOH]}{[CH_3COO^-]}$$

$$[H^+] = 1.76 \times 10^{-5} \times \frac{4.25 \times 10^{-3}}{2.00 \times 10^{-3}} = 3.74 \times 10^{-5}\,mol\,dm^{-3}$$

$$pH = -\log_{10}[H^+] = -\log_{10}(3.74 \times 10^{-5}) = 4.43$$

After equivalence

The excess of the strong base has to be found, together with the total volume of the solution. The concentration of the OH^- ions is then determined and the pH of the strong base is calculated using $K_w = [H^+][OH^-]$ and $pH = -\log_{10}[H^+]$.

Buffer solutions

A buffer is a solution that resists changes in pH when small amounts of acid or base are added. A buffer solution is also able to maintain its pH on dilution.

Acidic buffers

An acidic buffer maintains a solution at a pH below 7. It is a mixture of a weak acid and its conjugate base, for example ethanoic acid, CH_3COOH, and sodium ethanoate, CH_3COONa.

In this buffer solution there is:

- a large amount of undissociated ethanoic acid, because it is a weak acid
- a large amount of ethanoate ions, owing to the addition of sodium ethanoate which completely ionises in solution

This gives the following equilibrium:

$CH_3COOH(aq) \rightleftharpoons CH_3COO^-(aq) + H^+(aq)$
large amount   large amount

- On addition of a small amount of acid, the equilibrium will be driven to the left-hand side to remove the H^+ ions, thus maintaining the pH.
- On addition of a small amount of alkali, the OH^- ions will react with the H^+ ions to form H_2O. The equilibrium will be driven to the right-hand side to replace the H^+ ions, thus maintaining the pH.

Basic buffers

A basic buffer maintains a solution at a pH above 7. It is a mixture of a weak base and its conjugate acid, for example ammonia, NH_3, and ammonium chloride, NH_4Cl.

In this buffer solution there is:

- a large amount of undissociated ammonia, because it is a weak base
- a large amount of ammonium ions, owing to the addition of ammonium chloride which completely ionises in solution

This gives the following equilibrium:

$NH_3(aq) + H_2O(l) \rightleftharpoons NH_4^+(aq) + OH^-(aq)$
large amount   large amount

- On addition of a small amount of acid, the H^+ ions will react with the OH^- ions to form H_2O. The equilibrium will be driven to the right-hand side to replace the OH^- ions, thus maintaining the pH.
- On addition of a small amount of alkali, the equilibrium will be driven to the left-hand side to remove the OH^- ions, thus maintaining the pH.

Dilution of a buffer solution

The expression for K_a for ethanoic acid is:

$$K_a = \frac{[CH_3COO^-][H^+]}{[CH_3COOH]}$$

Rearranging gives:

$$[H^+] = K_a \times \frac{[CH_3COOH]}{[CH_3COO^-]}$$

On dilution, the concentrations of CH_3COOH and CH_3COO^- will change by the same amount. The concentration of H^+ will not change, so the pH will remain constant.

Calculating the pH of an acidic buffer solution

Consider the buffer solution formed when ethanoic acid and sodium ethanoate are mixed together.

$CH_3COOH \rightleftharpoons CH_3COO^- + H^+$

Calculate the pH of a buffer solution made by mixing 15.0 cm^3 of 1.00 mol dm^{-3} solution of ethanoic acid ($K_a = 1.76 \times 10^{-5}$ mol dm^{-3}) with 18.0 cm^3 of 0.800 mol dm^{-3} solution of sodium ethanoate.

$$K_a = \frac{[CH_3COO^-][H^+]}{[CH_3COOH]}$$

$$[H^+] = K_a \times \frac{[CH_3COOH]}{[CH_3COO^-]}$$

$$\text{moles of ethanoate ions} = \frac{0.800 \times 18.0}{1000} = 1.44 \times 10^{-2}$$

$$\text{moles of ethanoic acid} = \frac{1.00 \times 15.0}{1000} = 1.50 \times 10^{-2}$$

$$[H^+] = 1.76 \times 10^{-5} \times \frac{1.50 \times 10^{-2}}{1.44 \times 10^{-2}} = 1.83 \times 10^{-5}$$

$$pH = -\log_{10}[H^+] = -\log_{10}(1.83 \times 10^{-5}) = 4.74$$

Note: since the ethanoic acid and the ethanoate ions exist together in the same overall volume (in this case 33.0 cm^3), the ratio of the concentrations is equal to the ratio of the number of moles.

Applications of buffer solutions

Buffers have widespread applications. Many, but not all, of these applications involve biological systems since enzyme action is pH dependent.

- The growth of bacterial cultures, e.g. for hospital tests, is only possible in buffered systems.
- The blood has a complex buffering system to maintain the pH between 7.35 and 7.45; therefore all intravenous drips must be buffered within a very narrow range.
- Food preserves, e.g. jams, require the addition of a buffer.
- Chemical processes such as electroplating and dyeing are controlled by buffers.
- pH meters are calibrated using buffer solutions.

Nomenclature and isomerism in organic chemistry

Naming organic compounds

The general rules for naming organic compounds were given in Unit 1 and were applied to the naming of compounds containing a range of functional groups in both Unit 1 and Unit 2. For this topic your application of these rules will be extended to include esters, acyl halides, acid anhydrides, amines and benzene compounds.

The table below includes the functional groups met in this unit.

Type of compound	Functional group	Suffix/ prefix	Examples
Amines	$-NH_2$	amino- or -amine	$H_2N-CH_2-C(=O)OH$ Aminoethanoic acid; H_3C-NH_2 Methylamine
Amides	$-C(=O)NH_2$	-amide	$H_3C-C(=O)NH_2$ Ethanamide
Nitriles	$-C\equiv N$	-nitrile	$H_3C-CH_2-C\equiv N$ Propanenitrile
Esters	$-C(=O)OR$	-oate	$H_3C-C(=O)O-CH_2-CH_3$ Ethyl ethanoate
Acyl halides	$-C(=O)X$ X = Cl, Br, I	-oyl halide	$H_3C-C(=O)Cl$ Ethanoyl chloride
Acid anhydrides	$-C(=O)-O-C(=O)-$	-oic anhydride	$H_3C-C(=O)-O-C(=O)-CH_3$ Ethanoic anhydride
Arenes (aromatic)	benzene ring	-benzene or phenyl-	Chlorobenzene (C_6H_5Cl); Methylbenzene ($C_6H_5CH_3$); Phenylamine ($C_6H_5NH_2$); Phenylethene ($C_6H_5CH=CH_2$)

Aromatic compounds containing benzene rings with only one functional group are straightforward to name, e.g. nitrobenzene or chlorobenzene. When a benzene ring has more than one functional group, the groups are numbered around the ring in such a way that the lowest possible numbers are used. The functional groups are listed in alphabetical order, together with their appropriate numbers.

4-chloro-3-methylbenzenecarboxylic acid

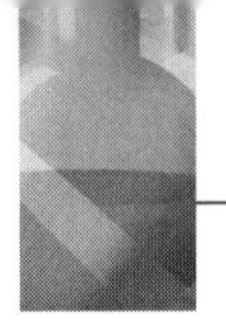

Isomerism

- You need to understand the basic principles of the two main types of isomerism, **structural isomerism** and **stereoisomerism**, know that each of these types may be further subdivided, and understand the basic principles of each sub-type. You need to be able to draw structural formulae and displayed formulae of isomers.
- You need to understand the meaning of the terms **asymmetric carbon atom**, **chiral**, **optical isomers**, **enantiomers** and **racemates**, and why racemates are formed.

Structural isomers

Structural isomerism occurs when there are two or more compounds with the same molecular formula but with different structural formulae.

Different types of structural isomerism include **chain isomerism**, **position isomerism** and **functional group isomerism**.

Structural isomerism was covered in more detail in Unit Guide 1.

Stereoisomerism

Stereoisomers are compounds that have the same molecular and structural formula but their bonds are arranged differently in space.

There are two types of stereoisomerism: ***E–Z*** **isomerism** (which was covered in Unit Guide 2) and **optical isomerism**.

Optical isomerism

Compounds that possess a carbon atom with four different groups attached to it tetrahedrally exhibit optical isomerism because the molecule is asymmetric, i.e. it has no centre, plane or axis of symmetry. Two tetrahedral arrangements in space are possible so that one is the mirror image of the other and cannot be superimposed on the other.

A molecule that is asymmetric and exhibits optical isomerism is said to be **chiral**, with the asymmetric carbon atom known as the **chiral centre**. The two mirror images are optical isomers known as **enantiomers**.

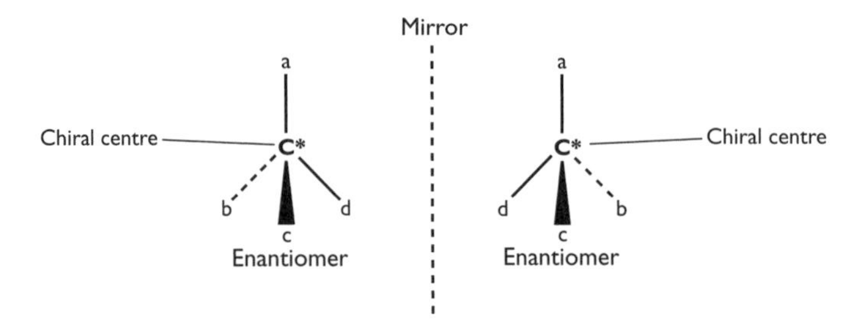

Enantiomers have the same molecular and structural formulae and differ only in their spatial arrangement. They have the same physical and chemical properties except they rotate the plane of plane-polarised light in opposite directions and are therefore said to be optically active. The enantiomer that rotates the plane of plane-polarised light in a clockwise direction is termed dextrorotatory and indicated by (+). Its mirror image will rotate the plane in an anti-clockwise direction by the same amount and is termed laevorotatory and indicated by (–). Mixing equal amounts of the two enantiomers gives an optically inactive mixture, which has no effect on plane-polarised light because the two effects cancel out. This mixture is known as a **racemic mixture** or **racemate**.

Many naturally occurring molecules exist as a single enantiomer, for example amino acids such as 2-aminopropanoic acid (alanine), $CH_3CH(NH_2)COOH$.

Another compound that exhibits optical isomerism is 2-hydroxypropanoic acid (lactic acid), $CH_3CH(OH)COOH$.

OH
H₃C—C*—C(=O)OH
H

The lactic acid found in sour milk is the racemate but it is the (+)-enantiomer that is formed during the contraction of muscles.

Enzymes in the body are stereospecific and distinguish between enantiomers and catalyse the reactions of only one of a pair of isomers.

Stereospecific drugs

The chemical properties of enantiomers are identical except in reactions with other optical isomers. Many commercially available drugs contain molecules with at least one chiral centre and those produced synthetically are obtained as racemic mixtures. Because of the difficulty and high cost of separating the enantiomers, such drugs have often been marketed as the racemic mixture. Drug action often depends on the action of one enantiomer in the body and the different enantiomers of the drug can act at quite different rates, e.g. the (+) isomer of the analgesic drug ibuprofen is faster acting than the (–) isomer. In some rare cases, the different enantiomers cause very different physiological reactions. For example, the (+) isomer of the drug Thalidomide affects the growth of a foetus and, when prescribed as the racemate as a sedative for pregnant women in the 1960s, caused terrible deformities; the (–) isomer has negligible side effects. Nowadays, the optical isomers of chiral drugs are isolated and tested separately.

Compounds containing the carbonyl group

Aldehydes and ketones

Nomenclature

Aldehydes and ketones are referred to as carbonyl compounds. They have the same general formula, $C_nH_{2n}O$, and contain the C=O group. Aldehydes have the C=O group at the end of the carbon chain; ketones have the C=O group elsewhere in the chain.

You need to be familiar with the naming of each type of carbonyl compound.

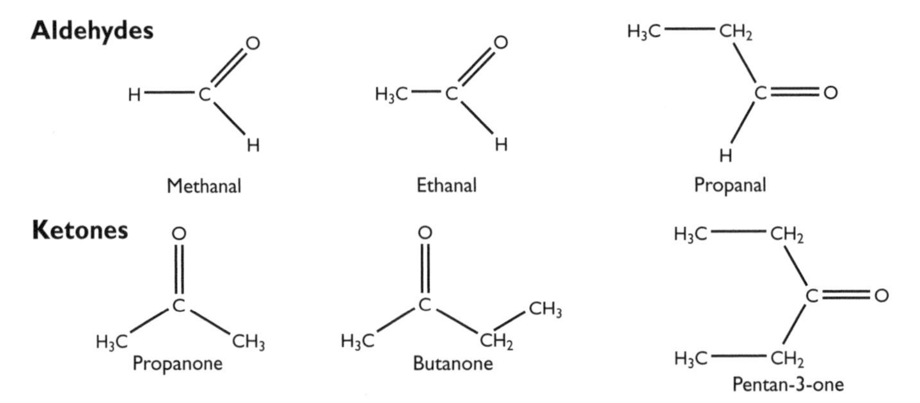

Distinguishing between aldehydes and ketones

The tests depend on the fact that aldehydes are stronger reducing agents than ketones. The aldehyde group possesses a hydrogen atom that can be oxidised easily to an –OH group, thus producing a carboxylic acid. Ketones are resistant to oxidation with mild oxidising agents.

Use of Tollens' reagent (the silver mirror test)

- **Procedure:** ammonia solution is added to aqueous silver nitrate until the brown precipitate of silver oxide dissolves, producing a colourless solution, which contains the complex ion $[Ag(NH_3)_2]^+$. This solution is added to the aldehyde and if there is no reaction, the mixture is gently warmed.
- **Observations:** the silver complex is reduced by an aldehyde to produce a metallic silver precipitate. A ketone gives no reaction.

Use of Fehling's solution

- **Procedure:** Fehling's solution is added to the aldehyde and warmed gently.
- **Observations:** the Cu^{2+} complex in Fehling's solution is reduced by an aldehyde to a red precipitate of copper(I) oxide, Cu_2O. A ketone gives no reaction.

Preparation of carbonyl compounds

The oxidation of primary alcohols produces aldehydes. For example:

$CH_3CH_2OH + [O] \longrightarrow CH_3CHO + H_2O$
Ethanol → Ethanal

The reagents needed are acidified potassium dichromate(VI) and dilute H_2SO_4. The low boiling point aldehyde is collected by distillation.

The oxidation of secondary alcohols produces ketones. For example:

$CH_3CH(OH)CH_3 + [O] \longrightarrow CH_3COCH_3 + H_2O$
Propan-2-ol → Propanone

The reagents needed are acidified potassium dichromate(VI) and dilute H_2SO_4. The reaction is carried out by heating the mixture under reflux.

Note: in equations for oxidation reactions in organic chemistry, [O] can be used to represent the oxidising agent.

The reduction of carbonyl compounds

Aldehydes and ketones are reduced using sodium tetrahydridoborate(III), $NaBH_4$. The reaction is carried out under reflux in aqueous ethanol (followed by acidification by adding dilute H_2SO_4).

- Aldehydes are reduced to primary alcohols. For example:
  $CH_3CH_2CH_2CHO + 2[H] \longrightarrow CH_3CH_2CH_2CH_2OH$
  butanal → butan-1-ol
- Ketones are reduced to secondary alcohols. For example:
  $CH_3CH_2COCH_3 + 2[H] \longrightarrow CH_3CH_2CH(OH)CH_3$
  butanone → butan-2-ol

Note 1: in equations for reduction reactions in organic chemistry, [H] can be used to represent reducing agents (other than hydrogen when H_2 must be used).

Note 2: if a compound contains a carbonyl group (C=O) and an alkene group (C=C), using $NaBH_4$ will reduce the carbonyl group only. Both groups will be reduced using hydrogen in the presence of a nickel catalyst. For example:

$CH_3CH{=}CHCHO + 2[H] \longrightarrow CH_3CH{=}CHCH_2OH$
$CH_3CH{=}CHCHO + 2H_2 \longrightarrow CH_3CH_2CH_2CH_2OH$

Nucleophilic addition in carbonyl compounds

Carbonyl compounds are unsaturated and undergo addition reactions. The C=O bond is polar because the oxygen atom is more electronegative than the carbon atom. The carbon atom of the C=O bond is electron deficient and is susceptible to attack by nucleophiles (electron-pair donors).

Example 1: addition of hydride ions (from $NaBH_4$) to ethanal

- **Reagent:** $NaBH_4$, followed by dilute H_2SO_4
- **Nucleophile:** hydride ion, H^-

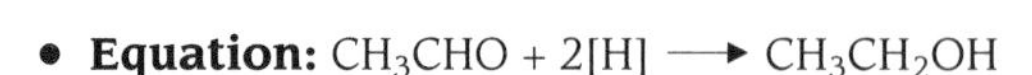

- **Equation:** $CH_3CHO + 2[H] \longrightarrow CH_3CH_2OH$
- **Product:** ethanol
- **Mechanism:**

Note: although [H] can be used in the equation, H^- must be used in the mechanism.

Example 2: addition of hydrogen cyanide (HCN) to ethanal

- **Reagent:** HCN (formed in the laboratory by adding a dilute acid to an excess of aqueous sodium cyanide)
- **Nucleophile:** cyanide ion, CN^-
- **Equation:** $CH_3CHO + HCN \longrightarrow CH_3CH(OH)CN$
- **Product:** 2-hydroxypropanenitrile (racemic mixture of two enantiomers)
- **Mechanism:**

All other carbonyl compounds react in a similar fashion to ethanal and undergo nucleophilic addition reactions with $NaBH_4$ and HCN. You need to be able to draw the mechanism for any carbonyl compound reacting with these reagents. The position of the curly arrows, to show electron pair movement, will always be the same. The only difference will be the structure of the carbonyl compound and the final product.

The use of HCN reactions in synthetic routes

When carbonyl compounds undergo nucleophilic addition reactions with HCN, the products formed contain the –CN group and the carbon chain of the original compound increases by an extra carbon atom. For example:

$CH_3CHO + HCN \longrightarrow CH_3CH(OH)CN$

ethanal → 2-hydroxy**prop**anenitrile

The –CN group can:

- be readily hydrolysed by boiling in dilute hydrochloric acid to produce a carboxylic acid group, –COOH:

  $CH_3CH(OH)CN + 2H_2O + HCl \longrightarrow CH_3CH(OH)COOH + NH_4Cl$

- be reduced by heating in hydrogen in the presence of a nickel catalyst to produce an amine, $–CH_2NH_2$:

  $CH_3CH(OH)CN + 2H_2 \longrightarrow CH_3CH(OH)CH_2NH_2$

Note: hydrogen cyanide is a highly toxic gas. Inhalation of hydrogen cyanide will cause rapid death. It is usual to generate HCN in a reaction mixture by adding a dilute acid to an excess of sodium/potassium cyanide. Care is required when handling sodium/potassium cyanide as these salts are also very toxic if ingested.

Carboxylic acids and esters

Carboxylic acids

Carboxylic acids are weak acids but they are strong enough to liberate CO_2 from carbonates. They have typical acid reactions with metals to form the salt and hydrogen, and with alkalis to form the salt and water. For example:

$$2CH_3COOH + Mg \longrightarrow (CH_3COO)_2Mg + H_2$$
$$CH_3COOH + NaOH \longrightarrow CH_3COONa + H_2O$$
$$2CH_3COOH + Na_2CO_3 \longrightarrow 2CH_3COONa + H_2O + CO_2$$

The structures and names of some carboxylic acids are given below.

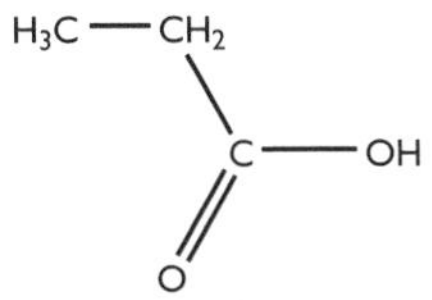

Propanoic acid

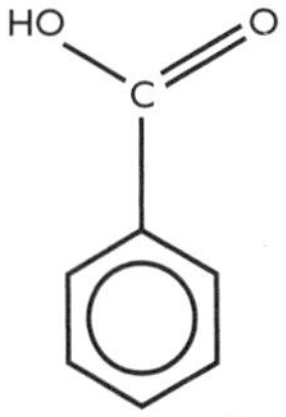

Benzenecarboxylic acid

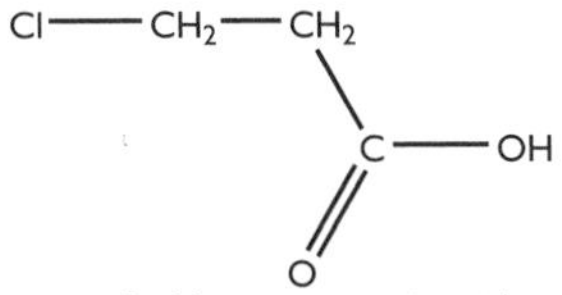

3-chloropropanoic acid

Carboxylic acids vary in strength, depending on the groups attached to the –COOH group. The acid strength is shown by the K_a value. The higher the K_a value, the greater is the dissociation of the acid, and the greater the strength of the acid.

Carboxylic acids react with alcohols in the presence of a strong acid catalyst to give esters.

Esters

Esters are formed by the reaction of a carboxylic acid with an alcohol. The reaction is:

$$\text{acid} + \text{alcohol} \rightleftharpoons \text{ester} + \text{water}$$

The acid and alcohol mixture is warmed in the presence of a strong acid catalyst, e.g. concentrated H_2SO_4.

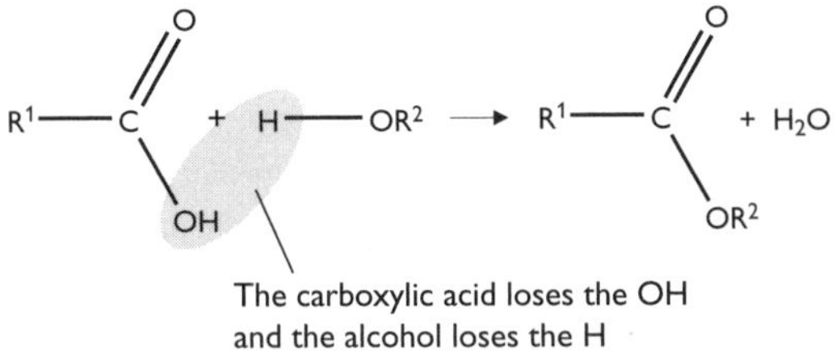

$$R^1COOH + H{-}OR^2 \longrightarrow R^1COOR^2 + H_2O$$

The carboxylic acid loses the OH and the alcohol loses the H

You should be able to predict the acids and alcohols used to prepare esters. For example, methyl ethanoate is prepared from methanol and ethanoic acid.

$$H_3C{-}OH + H_3C{-}C({=}O){-}OH \rightleftharpoons H_3C{-}C({=}O){-}O{-}CH_3 + H_2O$$

Methanol + Ethanoic acid ⇌ Methyl ethanoate + H_2O

You need to be able to name and predict the structure of the different esters produced in esterification reactions, know that esters have pleasant smells and be able to predict the products of ester hydrolysis.

The structures and names of some esters are given below.

$H{-}C({=}O){-}O{-}CH_3$ Methyl methanoate

$H{-}C({=}O){-}O{-}CH_2{-}CH_3$ Ethyl methanoate

$H_3C{-}CH_2{-}C({=}O){-}O{-}CH_3$ Methyl propanoate

Common uses of esters

- **Solvents:** esters dissolve many polar organic compounds and are volatile, so they are easily separated from the solute. For example, ethyl ethanoate is used in nail varnish, polystyrene cement and printing inks.
- **Plasticisers:** these are added to plastics to improve flexibility, for example esters of benzene-1,2-dicarboxylic acid (phthalic acid) are added to PVC. However, over time, the additives escape and the plastics become brittle.
- **Food flavourings:** many esters have sweet or fruity smells and are used as artificial fruit flavourings, for example pentyl ethanoate (pear) and octyl ethanoate (orange).
- **Perfumes:** when mixed with other compounds (such as alcohols); an example is pentyl 2-hydroxybenzenecarboxylate (jasmine).

Hydrolysis

The scheme below shows both the acidic and alkaline hydrolysis of esters.

Acid hydrolysis produces the acid and the alcohol

$$R^1{-}C({=}O){-}O{-}R^2 \xrightarrow[H_2O/HCl]{Heat} R^1{-}C({=}O){-}OH + R^2OH$$

$$R^1{-}C({=}O){-}O{-}R^2 \xrightarrow[NaOH]{Heat} R^1{-}C({=}O){-}O^-Na^+ + R^2OH$$

Alkaline hydrolysis produces the salt of the acid and the alcohol

The acid hydrolysis of ethyl methanoate is represented by:

$$HCOOCH_2CH_3 + H_2O \rightleftharpoons HCOOH + CH_3CH_2OH$$

The alkaline hydrolysis of propyl ethanoate is represented by:

$$CH_3COOCH_2CH_2CH_3 + NaOH \longrightarrow CH_3COONa + CH_3CH_2CH_2OH$$

The alkaline hydrolysis of animal fats and vegetable oils

Animal fats and vegetable oils are esters of propane-1,2,3-triol (glycerol) and long-chain carboxylic acids called fatty acids. Examples of fatty acids include octadecanoic acid (stearic acid), octadec-9-enoic acid (oleic acid) and octadeca-9,12-dienoic acid (linoleic acid). The esters formed between glycerol and these fatty acids are called **triglycerides**. Triglycerides can be hydrolysed by NaOH.

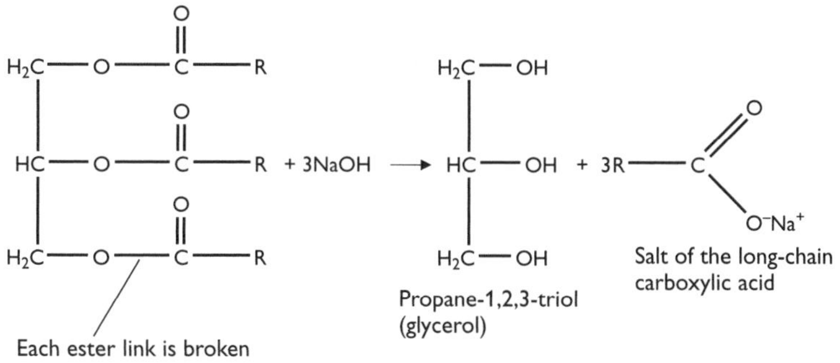

$$\begin{array}{l} H_2C-O-C(=O)-R \\ \;| \\ HC-O-C(=O)-R \\ \;| \\ H_2C-O-C(=O)-R \end{array} + 3NaOH \longrightarrow \begin{array}{l} H_2C-OH \\ \;| \\ HC-OH \\ \;| \\ H_2C-OH \end{array} + 3R-C(=O)O^-Na^+$$

Each ester link is broken

Propane-1,2,3-triol (glycerol)

Salt of the long-chain carboxylic acid

If the acid is octadecanoic acid, then $R = -(CH_2)_{16}CH_3$ and the salt produced is sodium octadecanoate (sodium stearate), $CH_3(CH_2)_{16}COONa$. The sodium salts of long-chain carboxylic acids obtained in this way are soaps.

Glycerol has three –OH groups and therefore exhibits extensive hydrogen bonding, so it has a high affinity for water. It is added to food and glues to prevent them drying out too quickly.

Biodiesel

Biodiesel is a mixture of methyl esters of long-chain carboxylic acids. It is obtained from vegetable oils such as those from soya bean and rapeseed by the process of **transesterification**. Transesterification is the reaction of an ester with an alcohol to produce a different ester and a different alcohol. The vegetable oil is converted into biodiesel by the reaction with methanol in the presence of an acid-based catalyst. For example:

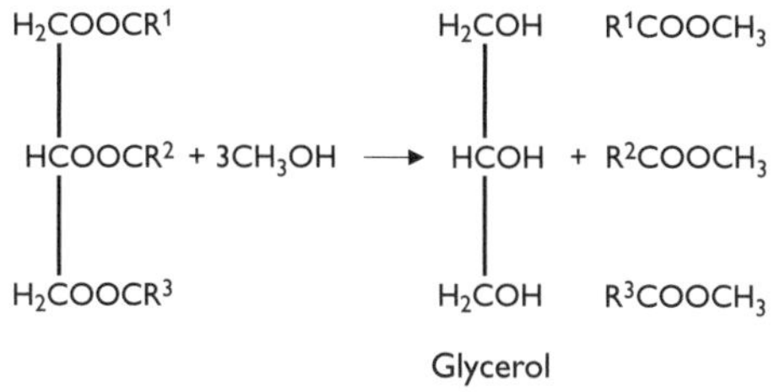

$$\begin{array}{l} H_2COOCR^1 \\ \;| \\ HCOOCR^2 \\ \;| \\ H_2COOCR^3 \end{array} + 3CH_3OH \longrightarrow \begin{array}{l} H_2COH \\ \;| \\ HCOH \\ \;| \\ H_2COH \end{array} + \begin{array}{l} R^1COOCH_3 \\ R^2COOCH_3 \\ R^3COOCH_3 \end{array}$$

Glycerol

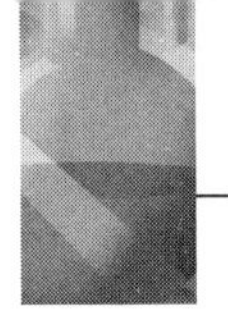

In the production of biodiesel the alkyl groups R^1, R^2 and R^3 can be the same or different. Propane-1,2,3-triol (glycerol) is always the other product.

Biodiesel can be used alone or blended with petrodiesel. It is a non-toxic, renewable, biodegradable fuel. As with bioethanol, there are concerns that land previously used for growing food will be turned over to growing crops for oils that can be used for converting into biodiesel. This could result in food shortages.

Acylation

Acylation reactions

Acylation involves the introduction of an **acyl** group into an organic compound. The reagents used in acylation are acyl chlorides and acid anhydrides.

R—C(=O)—
Acyl group

R—C(=O)—X
Acylating agent

In acyl chlorides X = Cl. In acid anhydrides X = OCOR.

CH_3COCl
Ethanoyl chloride

$(CH_3CO)_2O$
Ethanoic anhydride

Acyl chlorides and acid anhydrides show similar reactions, but the reactions with acid anhydrides are less vigorous. However, acid anhydrides produce co-products, which are not easily removed from the reaction mixture. Ethanoyl chloride and ethanoic anhydride are typical acylating agents and the equations for their reactions with water, alcohols, ammonia and primary amines are given below.

Example 1: with water to produce carboxylic acids

$$CH_3COCl + H_2O \longrightarrow CH_3COOH + HCl$$

Ethanoic acid

$$(CH_3CO)_2O + H_2O \longrightarrow 2CH_3COOH$$

Example 2: with alcohols to produce esters

$$CH_3COCl + CH_3OH \longrightarrow CH_3COOCH_3 + HCl$$

Methyl ethanoate

$$(CH_3CO)_2O + CH_3OH \longrightarrow CH_3COOCH_3 + CH_3COOH$$

The reaction of acyl chlorides with alcohols is a more effective method of preparing esters than esterification. Acylation is complete, whereas esterification is an equilibrium.

Example 3: with ammonia to produce amides

$$CH_3COCl + 2NH_3 \longrightarrow CH_3CONH_2 + NH_4^+Cl^-$$

Ethanamide

$$(CH_3CO)_2O + 2NH_3 \longrightarrow CH_3CONH_2 + CH_3COO^-NH_4^+$$

Ammonium ethanoate

Example 4: with amines to produce *N*-substituted amides

$$CH_3COCl + 2CH_3NH_2 \longrightarrow CH_3CONHCH_3 + CH_3NH_3^+Cl^-$$

N-methylethanamide

$$(CH_3CO)_2O + 2CH_3NH_2 \longrightarrow CH_3CONHCH_3 + CH_3COO^-CH_3NH_3^+$$

$$CH_3COCl + 2C_6H_5NH_2 \longrightarrow CH_3CONHC_6H_5 + C_6H_5NH_3^+Cl^-$$

N-phenylethanamide

Mechanism of nucleophilic addition–elimination

Both ethanoyl chloride and ethanoic anhydride undergo nucleophilic attack. The groups (Cl or CH_3COO) attached to the carbon in the C=O bond are strongly electron withdrawing, so they make the carbon electron deficient. The Cl^- ion and the CH_3COO^- group are stable and so act as good leaving groups. Typical nucleophiles are water, alcohols, ammonia and amines.

Example 1: ethanoyl chloride with water to form ethanoic acid

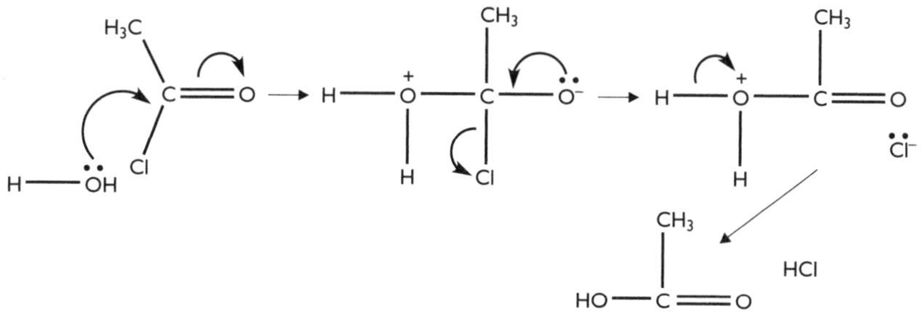

H₃C, C=O, Cl, H—ÖH → H—O⁺(H)—C(CH₃)(Cl)—O⁻ → H—O⁺(H)—C(CH₃)=O + Cl⁻ → HO—C(CH₃)=O + HCl

Example 2: ethanoyl chloride with ethanol to form ethyl ethanoate

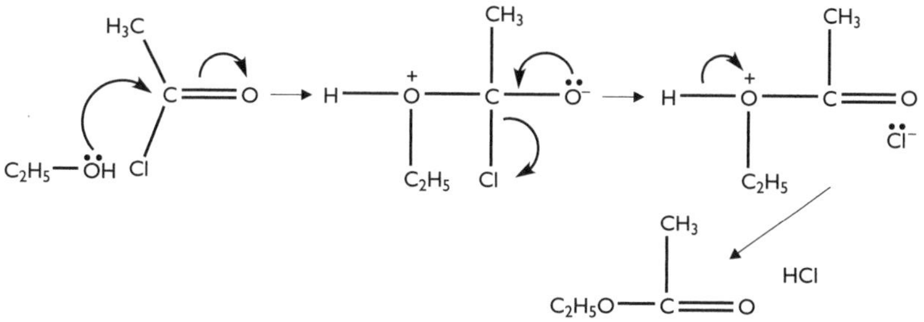

H₃C, C=O, Cl, C₂H₅—ÖH → H—O⁺(C₂H₅)—C(CH₃)(Cl)—O⁻ → H—O⁺(C₂H₅)—C(CH₃)=O + Cl⁻ → C₂H₅O—C(CH₃)=O + HCl

Example 3: ethanoyl chloride with ammonia to form ethanamide

Example 4: ethanoyl chloride with phenylamine to form *N*-phenylethanamide

Note: the equations for each of these reactions with ethanoic anhydride are required but the mechanisms are not.

Aspirin

Aspirin is a drug that is used widely, often as an analgesic (painkiller). There are two possible methods for its production. The –OH group in 2-hydroxybenzenecarboxylic acid undergoes acylation with either ethanoyl chloride or ethanoic anhydride.

The production of aspirin on a large scale is achieved using ethanoic anhydride as the acylating agent. The main advantages of ethanoic anhydride are that it is:

- cheaper than ethanoyl chloride
- less corrosive
- less vulnerable to hydrolysis
- less hazardous to use

Aromatic chemistry

Cyclic unsaturated compounds based on benzene are known as **arenes** but the term **aromatic** is still often used.

Benzene is a planar molecule with bond lengths intermediate between those of single and double bonds. You need to understand the unique nature of the bonding in the benzene ring, with its delocalised electrons, and that this delocalisation leads to increased stability in the benzene molecule.

The chemistry of arenes (aromatic compounds) is dominated by electrophilic substitution. You need to understand the mechanisms of nitration and Friedel–Crafts acylation.

The structure of benzene

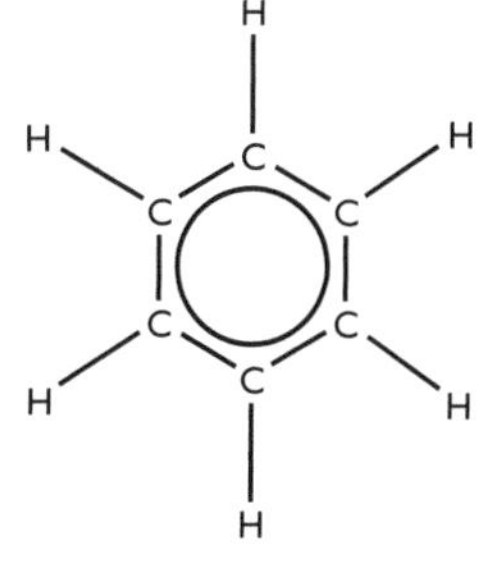

The key points are:

- It is a planar hexagonal molecule of six carbon atoms.
- Each carbon atom is bonded to two other carbon atoms and one hydrogen atom.
- All bond angles are equal at 120°.
- All carbon–carbon bond lengths are equal and are intermediate between a single C–C bond and a double C=C bond.
- Each carbon has one 'unused' *p* electron.
- Delocalisation of the six *p* electrons gives regions of electron density above and below the plane of the ring.
- The circle represents a ring of six delocalised electrons.

The stability of benzene

Benzene is more stable than expected owing to the presence of delocalised electrons. This can be shown by comparing the enthalpies of hydrogenation of cyclohexene, cyclohexa-1,3,5-triene and benzene.

Cyclohexene $+ H_2 \rightarrow$ cyclohexane $\Delta H = -120\ kJ\ mol^{-1}$

Cyclohexa-1,3,5-triene $+ 3H_2 \rightarrow$ cyclohexane $\Delta H = -360\ kJ\ mol^{-1}$

Benzene $+ 3H_2 \rightarrow$ cyclohexane $\Delta H = -208\ kJ\ mol^{-1}$

The possible structure of alternate single and double bonds for benzene would be expected to release three times the $-120\,kJ\,mol^{-1}$ of cyclohexene because it contains three times as many double bonds.

The enthalpy of hydrogenation of benzene is only $-208\,kJ\,mol^{-1}$; this is $152\,kJ\,mol^{-1}$ less than expected. Benzene requires the extra $152\,kJ\,mol^{-1}$ of energy to break the bonds; it is more stable than expected because of the ring of delocalised electrons. This extra stability is called the **delocalisation energy**.

Electrophilic substitution

Benzene, with its ring of delocalised electrons, is electron rich. This means that benzene is susceptible to attack by electrophiles, which are electron-pair acceptors.

The initial attack by the electrophile creates a positive intermediate, similar to the carbocation in the reaction of alkenes with electrophiles. However, this intermediate loses a proton in the second stage of the mechanism and the ring of delocalised electrons re-forms. Benzene undergoes substitution rather than addition in order to retain the stability associated with the ring of delocalised electrons.

Nitration

- **Reagents:** concentrated nitric acid (HNO_3) and concentrated sulfuric acid (H_2SO_4)
- **Equation:** $C_6H_6 + HNO_3 \longrightarrow C_6H_5NO_2 + H_2O$
- **Formation of the electrophile:** the concentrated sulfuric acid acts as a strong acid and donates an H^+ ion to the nitric acid. This produces the nitronium ion, NO_2^+, which acts as an electrophile.
  $HNO_3 + H_2SO_4 \rightleftharpoons H_2NO_3^+ + HSO_4^-$
  $H_2NO_3^+ \rightleftharpoons H_2O + NO_2^+$
  $H_2SO_4 + H_2O \rightleftharpoons HSO_4^- + H_3O^+$
  Overall: $HNO_3 + 2H_2SO_4 \rightleftharpoons NO_2^+ + 2HSO_4^- + H_3O^+$
- Mechanism:

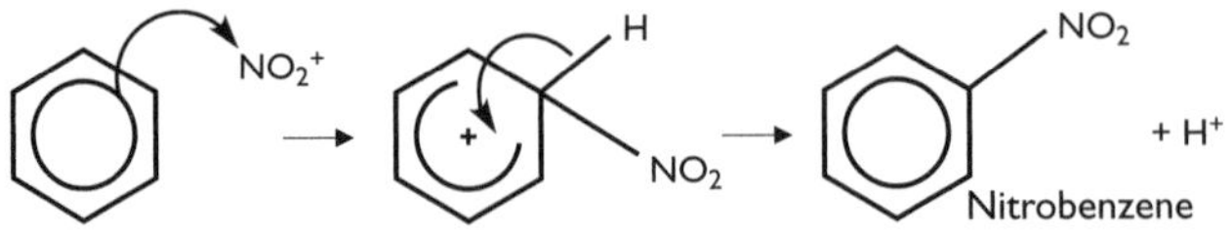

NO_2^+ → H, NO_2, + → NO_2 + H^+ Nitrobenzene

Importance of nitration

Polynitro compounds are unstable and decompose explosively to produce stable nitrogen, so they are used as explosives. The nitration of methylbenzene (toluene) produces a poly-substituted product, 2,4,6-trinitromethylbenzene, which is better known as 2,4,6-trinitrotoluene (TNT).

Nitrocompounds (containing $-NO_2$) can easily be reduced to amines (containing $-NH_2$). Amines can be converted into diazonium salts, which are then used in the production of dyes.

Friedel–Crafts acylation

- **Reagents:** an acyl chloride (e.g. CH_3COCl) and anhydrous aluminium chloride ($AlCl_3$)
- **Equation:** $C_6H_6 + CH_3COCl \longrightarrow C_6H_5COCH_3 + HCl$
- **Formation of the electrophile:** the aluminium chloride acts as a Lewis acid catalyst and accepts an electron pair from the chlorine, forming $AlCl_4^-$ and the electrophile CH_3CO^+:

  $CH_3COCl + AlCl_3 \longrightarrow CH_3CO^+ + AlCl_4^-$
- **Mechanism:**

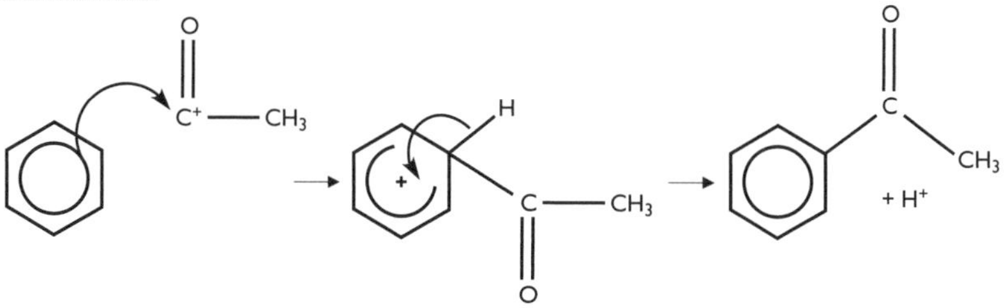

O, C⁺ — CH₃, H, C — CH₃, O, O, C, CH₃, + H⁺

Importance of Friedel–Crafts acylation

Friedel–Crafts reactions are important in synthesis as they lead to C–C bond formation. Acylation introduces a reactive (carbonyl) functional group to the ring. This can undergo the various reactions of a carbonyl group and so act as an intermediate in the synthesis of other compounds.

Note: Friedel–Crafts reactions involving halogenoalkanes (alkylation), rather than acyl halides, are covered in most textbooks but are not required in the AQA specification.

Amines

You need to understand the essential features of Brønsted–Lowry bases and be able to use these features to explain the differences in the base strength of ammonia, primary aliphatic amines (e.g. ethylamine) and primary aromatic amines (e.g. phenylamine) in terms of the availability of the lone pair on the nitrogen atom.

The chemistry of amines is dominated by their ability to act as nucleophiles due to the presence of the lone pair on the nitrogen atom.

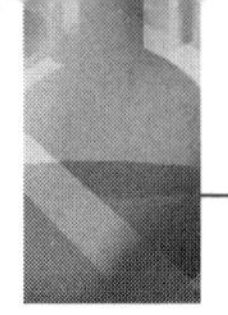

Basic properties

A Brønsted–Lowry base is defined as a proton acceptor. Ammonia (NH_3) and amines (RNH_2) have a lone pair of electrons on the nitrogen, which can accept a proton. The reactions of ammonia and amines with water and with an acid illustrate this.

$NH_3 + H_2O \rightleftharpoons NH_4^+ + OH^-$

$NH_3 + HCl \rightleftharpoons NH_4^+ + Cl^-$

$RNH_2 + H_2O \rightleftharpoons RNH_3^+ + OH^-$

$RNH_2 + HCl \rightleftharpoons RNH_3^+ + Cl^-$

The basicity (base strength) depends on the availability of the lone pair on the nitrogen atom. Compare ammonia, ethylamine and phenylamine:

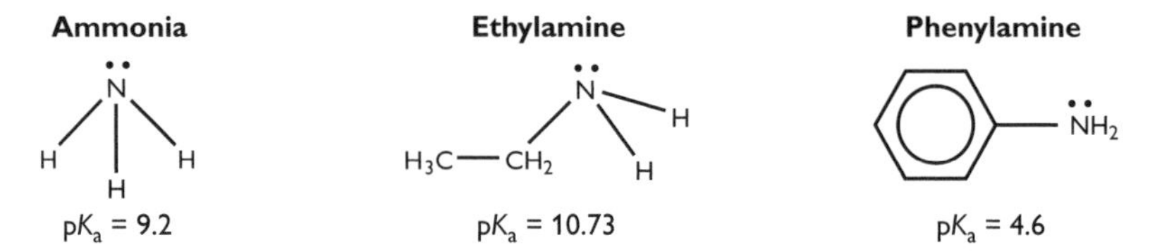

- Ammonia is used as the standard because it has no other groups attached.
- Ethylamine is a stronger base than ammonia because the lone pair on the nitrogen is more available due to the electron-releasing effect of the $-C_2H_5$ group.
- Phenylamine is a weaker base than ammonia because the lone pair is less available due to interaction with the delocalised electrons in the benzene ring.

The strength of a weak base is often indicated by the K_a or pK_a value of its conjugate acid; the stronger the conjugate acid the weaker the base. For example, ammonia ($pK_a = 9.2$) is a stronger base than phenylamine ($pK_a = 4.6$).

Note: many textbooks express the strength of a weak base in terms of the base dissociation constant K_b, which is analogous to the acid dissociation constant K_a. The larger the K_b value, the more protonation takes place and the stronger the base. The strength of the base can also be expressed in terms of pK_b. The AQA specification does not require you to cover K_b and pK_b.

Nucleophilic properties

A nucleophile is an electron-pair donor that attacks an electron-deficient site. Ammonia and amines have a lone pair of electrons on the nitrogen, so they act as nucleophiles and attack haloalkanes to give a mixture of products. For example:

$CH_3CH_2Br + NH_3 \longrightarrow CH_3CH_2NH_3^+Br^-$

$CH_3CH_2NH_3^+Br^- + NH_3 \longrightarrow CH_3CH_2NH_2 + NH_4Br$

primary amine

This is a nucleophilic substitution reaction and the initial product is ethylamine — a primary amine. If excess ammonia is used, the major product is ethylamine.

If excess bromoethane is used, further substitution reactions take place to produce a mixture of products — secondary and tertiary amines and quaternary ammonium salts. Each amine can continue to attack the haloalkane because the amine still has a lone pair on the nitrogen atom. If a large excess of bromoethane is used, a high yield of the quaternary ammonium salt is produced.

$$CH_3CH_2Br + CH_3CH_2NH_2 \longrightarrow (CH_3CH_2)_2NH_2^+Br^-$$
$$(CH_3CH_2)_2NH_2^+Br^- + NH_3 \longrightarrow \underset{\text{secondary amine}}{(CH_3CH_2)_2NH} + NH_4Br$$
$$CH_3CH_2Br + (CH_3CH_2)_2NH \longrightarrow (CH_3CH_2)_3NH^+Br^-$$
$$(CH_3CH_2)_3NH^+Br^- + NH_3 \longrightarrow \underset{\text{tertiary amine}}{(CH_3CH_2)_3N} + NH_4Br$$
$$CH_3CH_2Br + (CH_3CH_2)_3N \longrightarrow \underset{\text{quaternary ammonium salt}}{(CH_3CH_2)_4N^+Br^-}$$

The mechanism for the nucleophilic substitution reaction of primary haloalkanes with ammonia was explained in Unit Guide 2.

Quaternary ammonium salts are used in the production of cationic detergents, which are found in fabric softeners. The cationic detergent is added to the final rinse, after using anionic detergents, to impart a softer feel to the fabric. A typical cationic detergent contains two long-chain alkyl groups, for example $[CH_3(CH_2)_{17}]_2N(CH_3)_2^+Cl^-$.

Preparation of amines

Reaction of ammonia with haloalkanes

- **Reagents:** haloalkane (e.g. bromoethane) and *excess* ammonia in ethanol
- **Equation:** $CH_3CH_2Br + 2NH_3 \longrightarrow CH_3CH_2NH_2 + NH_4^+Br^-$

The reduction of nitriles

- **Reagent:** nitrile (e.g. propanenitrile), hydrogen and a nickel catalyst *or* $LiAlH_4$ in dry ether
- **Equations:**

$$CH_3CH_2CN + 2H_2 \longrightarrow CH_3CH_2CH_2NH_2$$
$$CH_3CH_2CN + 4[H] \longrightarrow CH_3CH_2CH_2NH_2$$

This route has two major advantages over the preparation of amines from haloalkanes — it gives a much better yield and there are no other products.

Reduction of aromatic nitro compounds

- **Reagents:** nitrobenzene, hydrogen and a nickel catalyst *or* tin and concentrated hydrochloric acid
- **Equations:**

$$C_6H_5NO_2 + 3H_2 \longrightarrow C_6H_5NH_2 + 2H_2O$$
$$C_6H_5NO_2 + 6[H] \longrightarrow C_6H_5NH_2 + 2H_2O$$

Amino acids

Amino acids have both acidic and basic properties due to the presence of the –COOH group and the $-NH_2$ group. They join together by peptide links to produce proteins, which can undergo hydrolysis at the peptide links to produce the constituent amino acids. You need to understand the importance of hydrogen bonding in proteins.

Structure

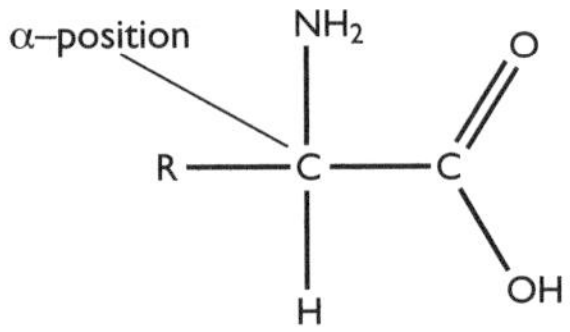

α–position NH_2 O R C C H OH

Amino acids contain a primary amino group ($-NH_2$) and a carboxyl group (–COOH). The primary amino group is attached to the carbon atom adjacent to the –COOH group, which is called the α-carbon. The molecules are referred to as α-amino carboxylic acids. There are 20 naturally occurring amino acids, which differ in the nature of the R group (see the structure above).

- The simplest has R = H; this is aminoethanoic acid (glycine), which is not optically active.
- All the other amino acids are optically active because they have four different groups attached to the α-carbon.

Amino acids are usually called by their common or trivial names.

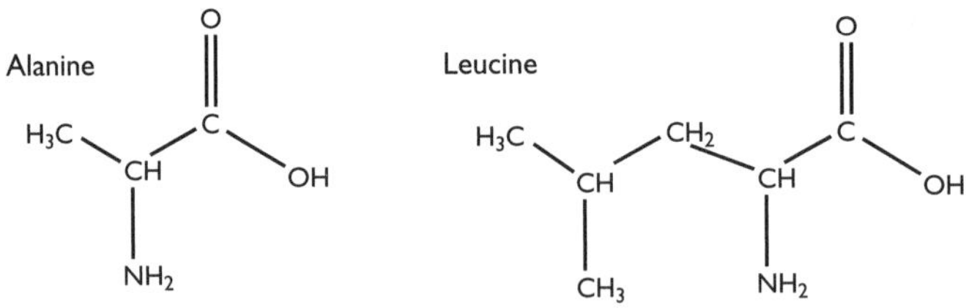

Alanine: H_3C, CH, NH_2, C, O, OH

Leucine: H_3C, CH, CH_3, CH_2, CH, NH_2, C, O, OH

Acid and base properties

Acid and base properties are demonstrated by all amino acids. This is illustrated below using aminoethanoic acid (glycine) as an example.

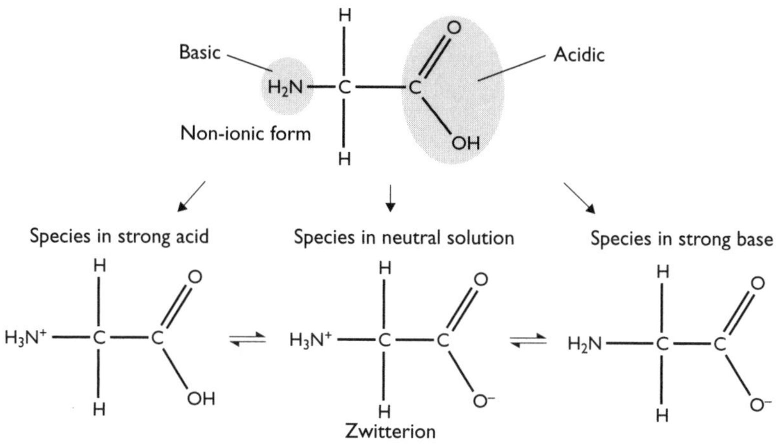
Basic
Acidic
H
H₂N—C—C
O
OH
H
Non-ionic form
Species in strong acid
Species in neutral solution
Species in strong base
H₃N⁺
H₂N
OH
O⁻
Zwitterion

Isoelectric point

The isoelectric point of an amino acid is the pH at which it has no net charge. It varies, depending on the R group. At the isoelectric point, an amino acid exists as a dipolar ion, $RCH(NH_3^+)COO^-$, called a **zwitterion**, as shown above.

The ionic nature of amino acids explains why they have such high melting points (e.g. glycine, 290°C).

Proteins

Amino acids link together to form polymers (called polypeptides and proteins). They link together through the formation of an amide group (–CONH–), which is called a **peptide link**.

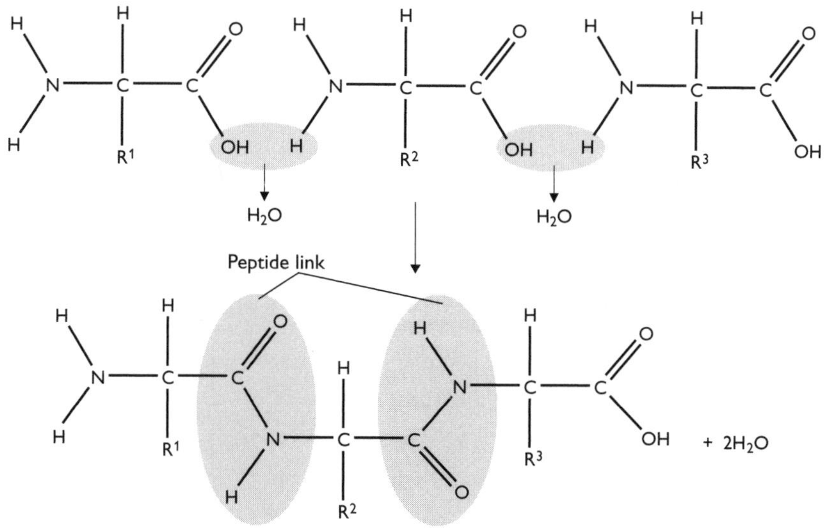
H
N
C
O
OH
R¹
R²
R³
H₂O
Peptide link
+ 2H₂O

The polymers show a range of relative molecular masses, for example:

- insulin, $M_r = 5700$, contains 51 amino acid residues
- haemoglobin, $M_r = 66\,000$, contains 574 amino acid residues

Hydrolysis of proteins

Proteins and polypeptides can be hydrolysed into their constituent amino acids. Hydrolysis is caused by reaction with an acid or a specific enzyme.

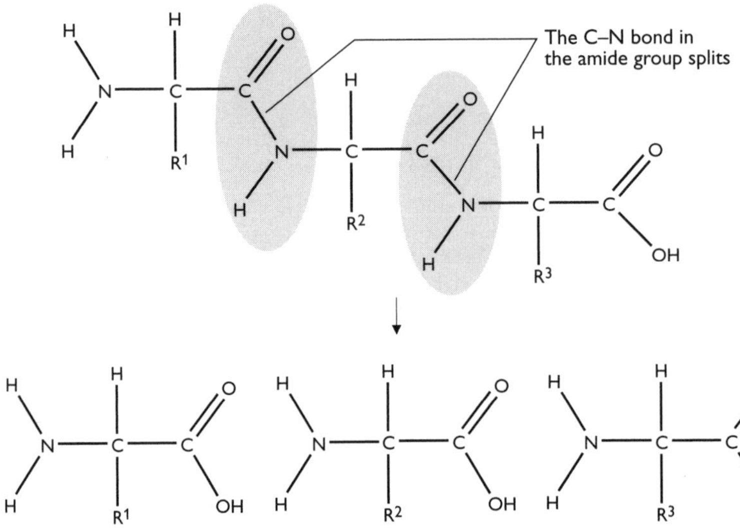

The C–N bond in the amide group splits

Mixtures of amino acids, such as those formed by the hydrolysis of a protein, can be separated by paper or thin-layer **chromatography**. As the amino acids formed are colourless, the chromatogram has to be sprayed with a chemical such as ninhydrin to reveal the positions of the spots so that the amino acids can be identified by their R_f values.

Hydrogen bonding in proteins

- The **primary** structure of a protein is the sequence of amino acids.
- The **secondary** structure of a protein is governed by hydrogen bonding, which results in the formation of helices or sheets. The hydrogen bonds occur between the atoms of the peptide links, in particular between oxygen atoms (from carbonyl groups) and hydrogen atoms (from amide groups).

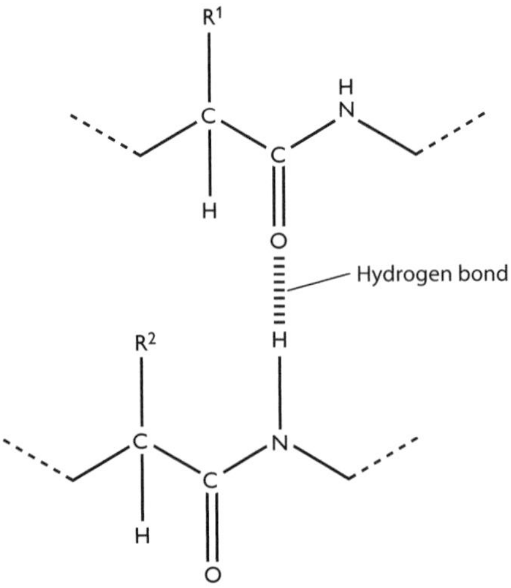

A large number of hydrogen bonds increases the forces within and between the proteins and forms stable structures.

Hydrogen bond

Polymers

Addition polymers

Addition polymers are formed directly from compounds containing C=C bonds. The unsaturated molecules that add together are called **monomers**. The long chain of monomers linked together is called a **polymer**. The addition polymerisation process can be represented by:

$$n\,CH_2{=}CHR \;\longrightarrow\; \left[\!-CH_2-CHR-\!\right]_n$$

Monomer → Polymer

R represents groups such as H, CH_3, C_6H_5, Cl, CN or $OCOCH_3$. The repeating unit in the polymer is enclosed in square brackets and n is a number between 100 and 10000.

Example 1: ethene to poly(ethene) or polythene

$$n\,H_2C{=}CH_2 \;\longrightarrow\; \left[\!-CH_2-CH_2-\!\right]_n$$

Example 2: phenylethene to poly(phenylethene) or polystyrene

$$n\,H_2C{=}CH(C_6H_5) \;\longrightarrow\; \left[\!-CH_2-CH(C_6H_5)-\!\right]_n$$

Condensation polymers

Condensation polymers are formed by the reaction between molecules with two functional groups and involve the loss of small molecules such as H_2O, HCl or CH_3OH.

- The reaction between a dicarboxylic acid and a diol leads to the formation of a polyester.
- The reaction between a dicarboxylic acid and a diamine leads to the formation of a polyamide.
- Amino acids can also be polymerised to form polyamides.

Example 1: the formation of Terylene, a polyester

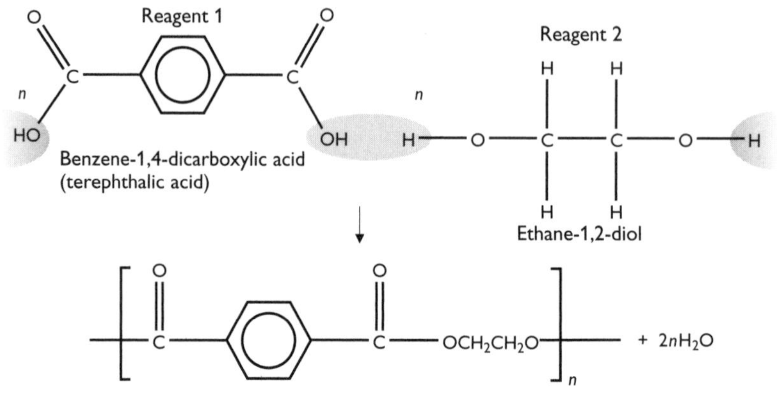

Reagent 1

Benzene-1,4-dicarboxylic acid (terephthalic acid)

Reagent 2

Ethane-1,2-diol

$$n\,HOOC{-}C_6H_4{-}COOH + n\,HO{-}CH_2CH_2{-}OH \rightarrow [{-}CO{-}C_6H_4{-}CO{-}OCH_2CH_2O{-}]_n + 2nH_2O$$

Example 2: the formation of nylon-6,6, a polyamide

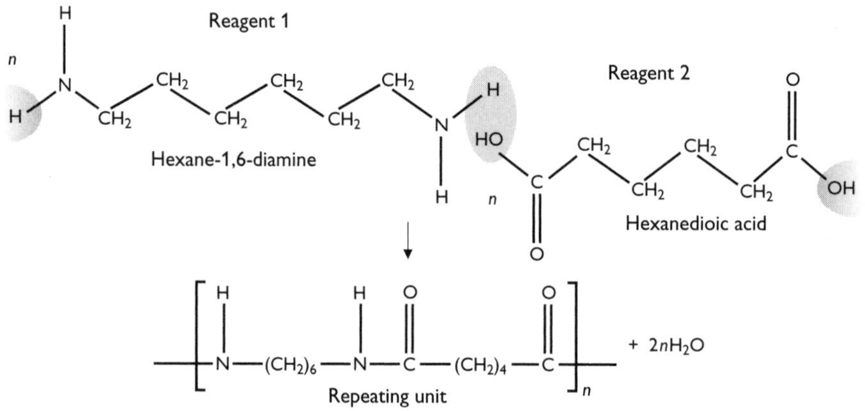

Reagent 1

Hexane-1,6-diamine

Reagent 2

Hexanedioic acid

$$n\,H_2N{-}(CH_2)_6{-}NH_2 + n\,HOOC{-}(CH_2)_4{-}COOH \rightarrow [{-}NH{-}(CH_2)_6{-}NH{-}CO{-}(CH_2)_4{-}CO{-}]_n + 2nH_2O$$

Repeating unit

Example 3: the formation of nylon-6, a polyamide

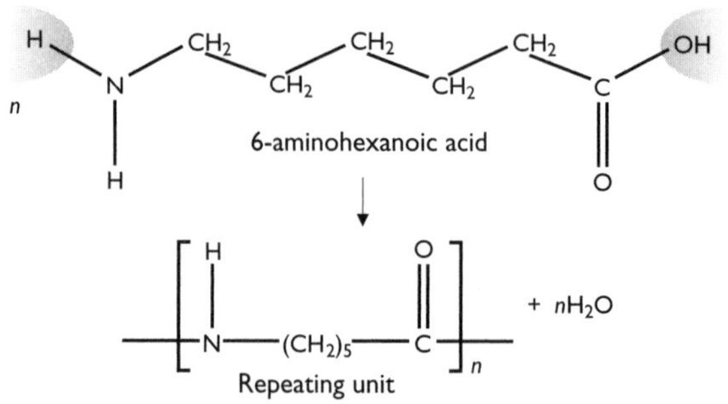

6-aminohexanoic acid

$$n\,H_2N{-}(CH_2)_5{-}COOH \rightarrow [{-}NH{-}(CH_2)_5{-}CO{-}]_n + nH_2O$$

Repeating unit

Note: there is a range of polyamides, referred to as 'nylons', each of which is identified by the number(s) at the end of their names. The numbers represent the number of carbon atoms in each monomer.

Another important polyamide is 'Kevlar', which forms fibres of great strength, stability and fire-resistance. It is used in applications such as bullet-proof vests. Kevlar is formed by the polymerisation reaction between benzene-1,4-dicarboxylic acid, $HOOC(C_6H_4)COOH$, and benzene-1,4-diamine, $H_2N(C_6H_4)NH_2$. The repeating unit of Kevlar is:

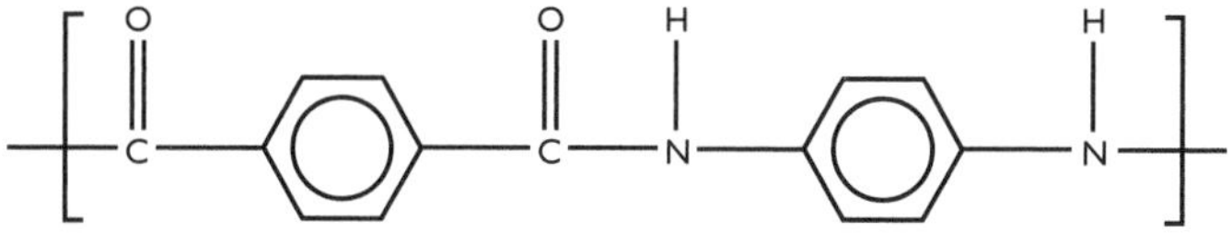

As with proteins, there is extensive hydrogen bonding between the amide groups of separate polyamide chains, which accounts for the great strength of polyamide fibres.

Biodegradability and disposal of polymers

Although addition polymers are named as polyalkenes, they are saturated molecules with C–C single bonds. They have no bond polarity so they are chemically inert and are non-biodegradable (not broken down by living organisms). Polymers such as poly(ethene) and poly(propene) have been used widely. Their disposal presents a considerable environmental problem; over the years most polymer waste has gone into rubbish tips (landfill sites) where it accumulates. As polymers are not biodegradable they will remain in the environment almost forever. In recent years biodegradable polymers have been developed and their availability and use will increase gradually. Biodegradable polymers must last long enough for the normal use of a product but must ultimately degrade to harmless products in landfill sites.

Polyamides and polyesters can undergo hydrolysis and the polymer chains are broken down into their component monomers. Enzyme-catalysed hydrolysis will occur in the environment, e.g. in landfill sites, which means they are biodegradable.

Polymers are flammable and another method for their disposal is to burn (incinerate) them. When polymers burn, toxic gases may be formed, such as carbon monoxide, hydrogen chloride (from polymers such as PVC containing chlorine), or possibly hydrogen cyanide (from polymers containing nitrogen). Modern incinerator plants are designed to deal with the toxic waste gases and operate at a sufficiently high temperature to break down any organic toxins. One advantage of the incineration of household waste is that it results in a net energy yield, which can be used in power generation or heating. It also considerably reduces the amount of waste sent to landfill.

Some of the disposal problems can be overcome by recycling. Many of the polymers used for household items/packaging are thermoplastics. This means that they melt on heating gently, can be remoulded while molten and will become solid again when cooled. They can then be reused, although not usually for their original purpose. In addition to decreasing the amount being disposed of, recycling plays a very small part in conserving crude oil supplies. However, it must be remembered that fossil fuels will be used during the recycling processes and so the savings are minimal. A disadvantage of recycling is that different polymers have different characteristics and must be separated into the different types before being recycled; also sufficient of each type must be accumulated. Although many household items can be recycled, a high proportion is currently still discarded as rubbish.

Note: the various types of plastics used in household packaging are identified by a numerical code given on the item. For example, high-density polyethene items would have a recycling logo and the number 2.

Organic synthesis and analysis

You need to be able to use the organic reactions covered both in this unit and in Unit 2 to convert one functional group to another in a multi-step process.

You will be expected to look at a multi-step process. For each individual step, you could be asked to:

- state the **reagents** used in the conversion
- classify the **reaction type**
- name and outline the **mechanism**
- draw and name the structures of the **intermediates**
- predict the final **product**

Synthesis

The tables below summarise the conversion of one functional group to another.

Reactions of alkenes

Conversion	Formulae	Reagents and conditions	Reaction type/ mechanism
Alkene → haloalkane	$RCH{=}CH_2 \rightarrow RCHBrCH_3$	HBr	Electrophilic addition
Alkene → dihaloalkane	$RCH{=}CH_2 \rightarrow RCHBrCH_2Br$	Br_2 in an organic solvent	Electrophilic addition
Alkene → alcohol	$RCH{=}CH_2 \rightarrow RCH(OH)CH_3$	Concentrated H_2SO_4 followed by H_2O	Electrophilic addition followed by hydrolysis

Reactions of haloalkanes

Conversion	Formulae	Reagents and conditions	Reaction type/ mechanism
Haloalkane ⟶ nitrile	$RBr \longrightarrow RCN$	NaCN in ethanol and reflux	Nucleophilic substitution
Haloalkane ⟶ alcohol	$RBr \longrightarrow ROH$	NaOH in aqueous solvent and reflux	Nucleophilic substitution (hydrolysis)
Haloalkane ⟶ amine	$RBr \longrightarrow RNH_2$	NH_3 in ethanol in a sealed vessel	Nucleophilic substitution
Haloalkane ⟶ alkene	$RCH_2CH_2Br \longrightarrow RCH{=}CH_2$	NaOH in alcoholic solvent and reflux	Elimination

Reactions of alcohols

Conversion	Formulae	Reagents and conditions	Reaction type/ mechanism
Primary alcohol ⟶ aldehyde	$RCH_2OH \longrightarrow RCHO$	Acidified $K_2Cr_2O_7$, *distillation*	Oxidation
Primary alcohol ⟶ acid	$RCH_2OH \longrightarrow RCOOH$	Acidified $K_2Cr_2O_7$, *reflux*	Oxidation
Secondary alcohol ⟶ ketone	$R_1R_2CHOH \longrightarrow R_1R_2CO$	Acidified $K_2Cr_2O_7$	Oxidation
Alcohol ⟶ alkene	$RCH_2CH_2OH \longrightarrow RCH{=}CH_2$	Concentrated H_2SO_4, heat	Elimination (dehydration)
Alcohol ⟶ ester	$R_1OH \longrightarrow R_2COOR_1$	Carboxylic acid (R_2COOH), concentrated H_2SO_4 as catalyst, warm	Esterification

Reactions of carbonyl groups

Conversion	Formulae	Reagents and conditions	Reaction type/ mechanism
Aldehyde ⟶ acid	$RCHO \longrightarrow RCOOH$	Acidified $K_2Cr_2O_7$, reflux	Oxidation
Aldehyde ⟶ primary alcohol	$RCHO \longrightarrow RCH_2OH$	Aqueous $NaBH_4$	Reduction
Ketone ⟶ secondary alcohol	$R_2CO \longrightarrow R_2CHOH$	Aqueous $NaBH_4$	Reduction
Aldehyde or ketone ⟶ hydroxynitrile	$RCHO \longrightarrow RCH(OH)CN$	HCN	Nucleophilic addition

Reactions of nitriles

Conversion	Formulae	Reagents and conditions	Reaction type/ mechanism
Nitrile ⟶ carboxylic acid	$RCN \longrightarrow RCOOH$	Dilute acid (HCl), reflux	Hydrolysis
Nitrile ⟶ amine	$RCN \longrightarrow RCH_2NH_2$	H_2, Ni catalyst, heat or $LiAlH_4$ in dry ether	Reduction

Reactions of esters

Conversion	Formulae	Reagents and conditions	Reaction type/ mechanism
Ester ⟶ salt of acid + alcohol	$R_1COOR_2 \longrightarrow R_1COONa + R_2OH$	NaOH, reflux	Alkaline hydrolysis
Ester ⟶ acid + alcohol	$R_1COOR_2 \longrightarrow R_1COOH + R_2OH$	HCl, reflux	Acid hydrolysis

Reactions of acyl chlorides

Conversion	Formulae	Reagents and conditions	Reaction type/mechanism
Acyl chloride ⟶ acid	$RCOCl \longrightarrow RCOOH$	H_2O	Nucleophilic addition–elimination (acylation)
Acyl chloride ⟶ ester	$R_1COCl \longrightarrow R_1COOR_2$	Alcohol (R_2OH)	Nucleophilic addition–elimination (acylation)
Acyl chloride ⟶ amide	$RCOCl \longrightarrow RCONH_2$	NH_3	Nucleophilic addition–elimination (acylation)
Acyl chloride ⟶ *N*-substituted amide	$R_1COCl \longrightarrow R_1CONHR_2$	Amine (R_2NH_2)	Nucleophilic addition–elimination (acylation)

Reactions of arenes (aromatic compounds)

Conversion	Formulae	Reagents and conditions	Reaction type/ mechanism
Benzene ⟶ nitrobenzene	$C_6H_6 \longrightarrow C_6H_5NO_2$	Concentrated HNO_3 and concentrated H_2SO_4, reflux	Electrophilic substitution (nitration)
Nitrobenzene ⟶ phenylamine	$C_6H_5NO_2 \longrightarrow C_6H_5NH_2$	Tin, concentrated HCl, reflux	Reduction
Benzene ⟶ phenylethanone	$C_6H_6 \longrightarrow C_6H_5COCH_3$	CH_3COCl and $AlCl_3$	Electrophilic substitution (Friedel–Crafts acylation)

$NO_2 + 6[H] \rightarrow NH_2 + 2H_2O$

Worked examples

It is essential that you learn *all* the reagents, conditions, types of reactions and mechanisms outlined in the tables above. You may also be asked to write balanced equations for the conversions.

Example 1

Consider this reaction sequence.

$H_3C—CH_2—CH_2—OH$ (A) —Step 1→ $H_3C—CH_2—CH_2—Br$ (B) —Step 2→ $H_3C—CH_2—CH_2—C{\equiv}N$ (C) —Step 3→ $H_3C—CH_2—CH_2—COOH$ (D) —Step 4→ $H_3C—CH_2—CH_2—COO—CH_3$ (E)

(1) Name compounds A to E. (5 marks)
(2) Classify the reaction types for steps 2, 3 and 4. (3 marks)
(3) State the reagents and conditions needed for steps 2 and 4. (2 marks)

Answers

(1) A = propan-1-ol; B = 1-bromopropane; C = butanenitrile; D = butanoic acid; E = methyl butanoate

(2) Step 2 = nucleophilic substitution; step 3 = hydrolysis; step 4 = esterification

(3) Step 2 = NaCN in ethanol and reflux; step 4 = methanol, concentrated H_2SO_4 as catalyst, warm

Example 2

This reaction sequence shows the preparation of *N*-phenylethanamide from benzene.

Benzene —Step 1→ A ($C_6H_5NO_2$) —Step 2→ B ($C_6H_5NH_2$) —Step 3→ *N*-phenylethanamide ($C_6H_5NHCOCH_3$)

(1) Name the intermediates A and B. (2 marks)
(2) State the reagents used in step 1. (2 marks)
(3) State the name of the mechanism for step 1. (1 mark)

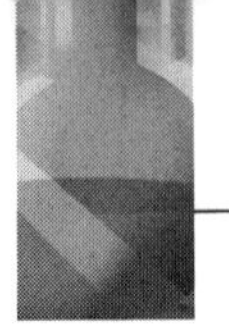

(4) State the type of reaction in step 2. (1 mark)
(5) State the reagents and conditions for step 2. (3 marks)
(6) State the name of the mechanism for step 3. (1 mark)

Answers

(1) A = nitrobenzene; B = phenylamine
(2) Concentrated nitric acid and concentrated sulfuric acid
(3) Electrophilic substitution
(4) Reduction
(5) Tin and concentrated hydrochloric acid; reflux
(6) Nucleophilic addition–elimination

Analysis

You need to be able to describe simple chemical tests that will distinguish between organic compounds containing different functional groups.

Alkenes

- **Test:** bromine water
- **Observation:** decolorises

Haloalkanes

- **Test:** warm with NaOH, acidify with HNO_3, then add $AgNO_3$
- **Observation:** chloroalkane gives a white precipitate of AgCl, bromoalkane gives a cream precipitate of AgBr and iodoalkane gives a yellow precipitate of AgI

Alcohols

- **Test:** acidified potassium dichromate(VI), $K_2Cr_2O_7$
- **Observation:** primary and secondary alcohols turn the potassium dichromate solution from orange to green; tertiary alcohols give no colour change

- **Test:** warm with ethanoic acid, CH_3COOH, in the presence of concentrated sulfuric acid
- **Observation:** smell of an ester

Aldehydes

- **Test:** warm with Fehling's solution
- **Observation:** the Fehling's solution changes from a blue solution to a red precipitate of copper(I) oxide, Cu_2O

- **Test:** warm with Tollens' reagent
- **Observation:** the colourless solution forms a silver mirror

- **Test:** acidified potassium dichromate(VI), $K_2Cr_2O_7$
- **Observation:** the potassium dichromate solution turns from orange to green

Note: the third test is also valid for primary and secondary alcohols, so an additional test may be necessary to confirm which of the functional groups is present.

Carboxylic acids

- **Test:** add aqueous sodium carbonate or sodium hydrogencarbonate solution
- **Observation:** fizzing due to evolution of carbon dioxide gas, CO_2

- **Test:** warm with ethanol, CH_3CH_2OH, in the presence of concentrated sulfuric acid
- **Observation:** smell of an ester

Acyl chlorides

- **Test:** add aqueous silver nitrate solution
- **Observation:** a vigorous reaction takes place to liberate HCl; the HCl immediately reacts with the silver nitrate to produce a white precipitate (of silver chloride)

Structure determination

You should be able to use data from each of the analytical techniques given below to determine the structure of specified compounds. You are not required to have a detailed knowledge of how such data are obtained.

Mass spectrometry

The determination of the molecular formula of a compound from the mass of the molecular ion was covered in Unit 2.

Fragmentation

Fragmentation of the molecular ion takes place to produce a characteristic pattern or stick diagram. The molecular ion has an unpaired electron, so it is a **radical cation**. When this molecular ion fragments, it produces a cation (which is detected) and a radical (which is undetected). The general equation for this fragmentation is as follows:

$$M^{+\bullet} \longrightarrow X^{+} + Y^{\bullet}$$

The most common ion gives rise to the **base peak**. By convention, the base peak is given a relative abundance of 100%, and all other peaks are expressed as a percentage of this.

Molecular ions fragment where bonds are weakest and dominant peaks are associated with stable cations such as carbocations (CH_3^+, $C_2H_5^+$) and acylium cations (CH_3CO^+, $C_2H_5CO^+$).

The mass spectrum of butanone

The mass spectrum of butanone is used to illustrate these principles.

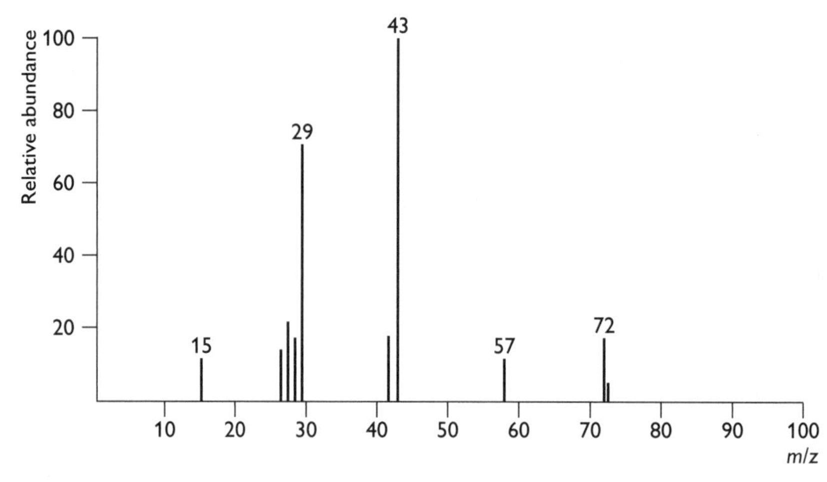

Butanone has the molecular formula C_4H_8O, so the molecular peak occurs at 72. The base peak is the most common peak and occurs at 43. Butanone has the structure shown. The arrows represent the bonds that break to produce the fragments at 15, 29, 43 and 57.

CH_3—C(=O)—CH_2—CH_3

Butanone

The four peaks are due to CH_3^+ at 15, $CH_3CH_2^+$ at 29, CH_3CO^+ at 43 and $CH_3CH_2CO^+$ at 57. The four fragments are formed from the parent ion as follows:

$$[CH_3COCH_2CH_3]^{+\bullet} \longrightarrow CH_3^+ + CH_3CH_2CO^\bullet$$
$$[CH_3COCH_2CH_3]^{+\bullet} \longrightarrow CH_3CH_2^+ + CH_3CO^\bullet$$
$$[CH_3COCH_2CH_3]^{+\bullet} \longrightarrow CH_3CO^+ + CH_3CH_2^\bullet$$
$$[CH_3COCH_2CH_3]^{+\bullet} \longrightarrow CH_3CH_2CO^+ + CH_3^\bullet$$

Infrared spectroscopy

The principles of infrared spectroscopy were covered in Unit 2. In this unit the application and use of spectra is extended to include all functional groups in the specification. As before you will always be given a table of relevant infrared absorptions on the examination paper.

Nuclear magnetic resonance spectroscopy

Basic principles

Nuclear magnetic resonance (NMR) occurs in atoms with odd-numbered nuclei, such as ^1H and ^{13}C.

The nuclei of atoms such as ^1H and ^{13}C have **nuclear spin** and possess a **magnetic moment**. This means that the nuclei behave like tiny bar magnets. When an external magnetic field is applied, the nucleus can have two spin states. It can align itself either with the external field (low-energy state) or against the field (high-energy state). A signal is recorded when a nucleus absorbs radiation in the radio-frequency range and resonates between the two spin states. Each nucleus in a unique environment within a molecule resonates at a specific frequency, resulting in characteristic peaks in the NMR spectrum.

^1H NMR spectra

^1H nuclei are protons. In an organic molecule, the protons are surrounded by the electrons in the molecule's covalent bonds. These electrons affect the NMR absorptions. The electrons are charged and have spin, and so have their own magnetic field, which can shield the proton from the external field. The amount of shielding depends on the electron density surrounding the nucleus, which varies for different protons in a molecule.

- The nucleus is **deshielded** when the electron density is reduced, owing to the presence of an electron-withdrawing group.
- The nucleus is **shielded** when the electron density is increased, owing to the presence of an electron-donating group.

Protons in different chemical environments give different peaks and chemically equivalent protons absorb at the same frequency. The movement of the signals caused by shielding (moving the signal upfield) and deshielding (moving the signal downfield) are measured by the **chemical shift, δ**. Chemical shifts are measured in parts per million (ppm) relative to an internal standard, tetramethylsilane, **TMS**.

The internal standard

CH_3
Si
H_3C CH_3 CH_3

Tetramethylsilane, TMS, is added to a sample being analysed to act as an internal standard:

- It gives a signal that resonates upfield from almost all other organic hydrogen resonances because the 12 equivalent hydrogens are highly shielded.
- It gives a single intense peak because there are 12 equivalent protons.
- It is non-toxic and inert.
- It has a low boiling point and can be easily removed from the sample being analysed.
- By definition, the δ value of TMS is zero and almost all proton NMR absorptions occur 0–10 ppm downfield from TMS.

The use of solvents

The sample being analysed must be dissolved in a solvent that is proton-free to avoid any unwanted absorptions. Typical solvents are CCl_4 and deuterated compounds such as $CDCl_3$ and C_6D_6, where D = ^2H.

Low-resolution spectra

The low-resolution spectrum of ethanol is shown below.

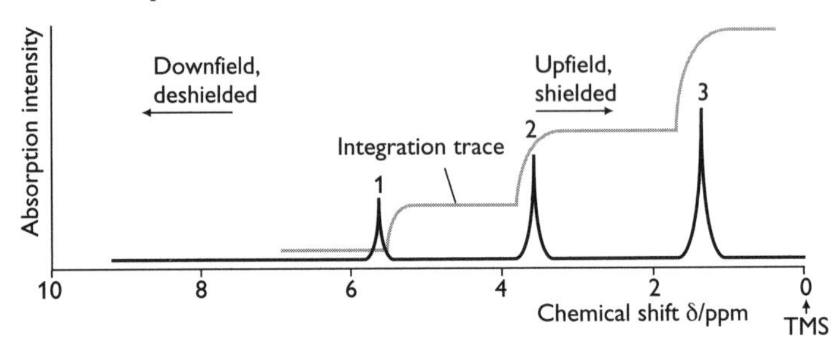

- The *number* of absorptions indicates the number of non-equivalent protons that are present. For example, ethanol has three non-equivalent protons.
- The *intensities* of the absorptions reveal how many protons are associated with that peak. In ethanol the absorptions are in the ratio 1:2:3, which are the –O**H**, the –C**H$_2$**– and –C**H$_3$** groups. The relative intensities of the various absorptions can be measured electronically to provide an integration trace. The height of each step in the trace is proportional to the number of equivalent hydrogen atoms.
- The *position* of the peaks gives information about the chemical environment of the protons. In ethanol, the closer the protons are to the electron-withdrawing oxygen atom the more deshielded they are and the further downfield they appear.

High-resolution spectra

The high-resolution spectrum of ethanol reveals that not all the absorptions are single peaks. The split peaks are shown here as lines to simplify the diagram.

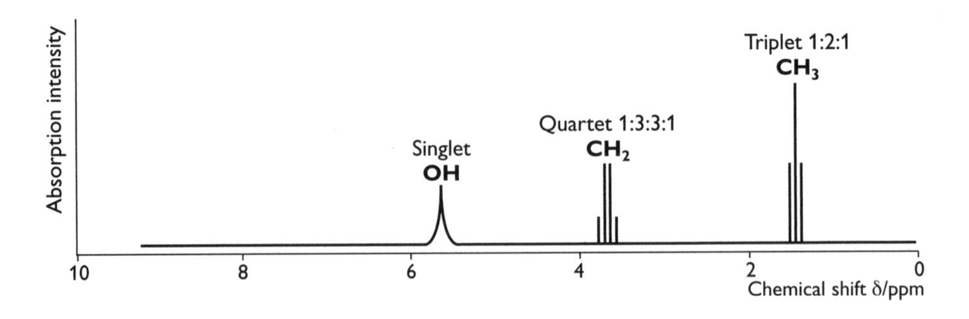

- The –OH peak still appears as a singlet.
- The $-CH_2-$ group appears as a quartet, that is, the signal is split into four peaks.
- The $-CH_3$ group appears as a triplet, that is, the signal is split into three peaks.

Non-equivalent hydrogen atoms on adjacent carbon atoms interact or **couple** with the protons, causing this splitting of peaks. This is called **spin–spin coupling**. The splitting of an absorption signal is described by the $n+1$ rule. Signals for protons adjacent to n equivalent neighbours split into $n+1$ peaks:

- The $-CH_3$ signal is split into a 1:2:1 triplet because it has two neighbouring protons.
- The $-CH_2-$ signal is split into a 1:3:3:1 quartet because it has three neighbouring protons.
- If there were a group with one neighbouring proton, then its signal would be split into a 1:1 doublet, but there is no such signal in ethanol.

The –OH proton appears as a singlet, devoid of any splitting. The –OH group is next to a $-CH_2-$ group, so spin–spin coupling should produce a triplet for the –OH signal. However, the –OH group is weakly acidic, so the proton ionises rapidly and transfers to other ethanol molecules and to water molecules. The –OH absorptions of this type are decoupled by this fast proton exchange and no splitting occurs.

The position of the –OH signal in an NMR spectrum can vary. The signal can be identified by the addition of D_2O: the –OH group becomes an –OD group and the –OH signal disappears from the spectrum.

Note: questions set will be limited to doublet, triplet and quartet formation in simple aliphatic compounds.

^{13}C NMR

As with protons, ^{13}C atoms in different chemical environments give different peaks and chemically equivalent ^{13}C atoms absorb at the same frequency.

- The *number* of absorptions indicates the number of non-equivalent carbon atoms that are present. For example, ethanol, CH_3CH_2OH, and methylpropan-2-ol, $(CH_3)_3COH$, both have two non-equivalent carbon atoms.
- The *position* of the peaks gives information about the different types of carbon atoms present in the compound.

As with ^1H NMR, chemical shift, δ, values are measured in parts per million (ppm) relative to the internal standard, tetramethylsilane, TMS. (The carbon atoms of the methyl groups are used for reference rather than the hydrogen atoms.) The range of chemical shifts in ^{13}C NMR spectroscopy is much wider than that for proton NMR and peaks are less likely to overlap.

The ^{13}C isotope has a low natural abundance (about 1.1%). The chance of finding more than one ^{13}C atom in a molecule is almost nil, and spin–spin coupling between adjacent ^{13}C atoms in a molecule is extremely unlikely. Although spin–spin coupling between ^{13}C and ^1H does exist, most ^{13}C spectra are obtained as proton-decoupled spectra. ^{13}C spectra are simpler than ^1H spectra and are generally easier to interpret.

Note 1: examination questions will only refer to proton-decoupled spectra with no splitting of the ^{13}C peaks. Relevant chemical shift data will be provided.

Note 2: unlike ^1H spectra, the *intensities* of the absorptions are not directly related to the number of carbon atoms associated with that peak.

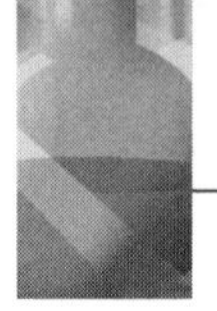

Chromatography

Chromatography is the term for a range of closely related techniques used to separate and identify mixtures. All involve a *mobile* (moving) phase moving through a *stationary* phase.

- **Adsorption chromatography** uses a *solid* stationary phase, e.g. thin-layer chromatography (TLC) and column chromatography.
- **Partition chromatography** uses a *liquid* stationary phase (held on the surface of an inert solid or within the fibres of a supporting material), e.g. paper chromatography and gas–liquid chromatography (GLC).

Separation is achieved because the components of the mixture are attracted to both phases to differing degrees and become distributed between them. The rate at which a component moves depends on its relative attraction to each phase. If a component is strongly attracted to the stationary phase, it will move slowly but one with a strong attraction to the mobile phase will move quickly. The phases must be carefully chosen to ensure separation.

Gas–liquid chromatography (GLC)

This type of chromatography can be used to separate mixtures of gases, of liquids and of volatile solids; it can be used for analysis of mixtures or in the industrial preparation of materials. It can be used to separate liquids with similar boiling points, e.g. benzene (b.p. 80.1°C) and cyclohexane (b.p. 80.8°C); these would be (almost) impossible to separate by fractional distillation. GLC can also be used to determine the purity of a compound.

The moving phase is a chemically inert gas of high purity, e.g. nitrogen, and is referred to as the **carrier gas**. The stationary phase is a non-volatile liquid coated onto an inert solid support and packed into a long, narrow-bore, coiled tube, the **column**. The sample is injected into the sample port where it vaporises and is carried through the column by the continuous flow of carrier gas. The components of a mixture move through the column at different rates, depending on their boiling points and their relative solubilities in the mobile and stationary phases. They are detected at the end of the column after definite intervals of time (after the initial injection of the sample) known as the **retention time**.

Information from the detector can be fed to a computer (and sometimes to a mass spectrometer). The chromatogram produced can be used to identify the components by comparing the retention times with those of known compounds obtained using the same stationary phase, the same carrier gas, the same flow rate and the same temperature. The area under the peak for each component in the chromatogram is related to the amount of that component in the mixture, i.e. GLC can provide both qualitative and quantitative information.

Column chromatography

In column chromatography the stationary phase is a finely divided absorbent solid, often alumina or silica gel, packed into a column (in a school laboratory this could be a burette). The sample to be separated is added to the top of the column in a minimum amount of solvent. The moving phase, fresh solvent referred to as the **eluent**, is run slowly through the column. Samples (fractions) of the solution coming from the column are collected periodically. Each fraction can be analysed for dissolved components and consecutive fractions containing the same component can be combined and evaporated to obtain the separated material. The time taken for a component to flow through a column is called the **retention time.** This depends on the balance between solubility in the moving phase and retention in the stationary phase. In general, the more polar the component, the more strongly it is attracted to the stationary phase and the greater the retention time.

Column chromatography can be used to separate and purify individual organic compounds from mixtures. High-performance liquid chromatography (HPLC) can be used to separate components that are very similar to each other. High pressure is used to force the liquid phase through the stationary phase, which is tightly packed into large stainless steel columns.

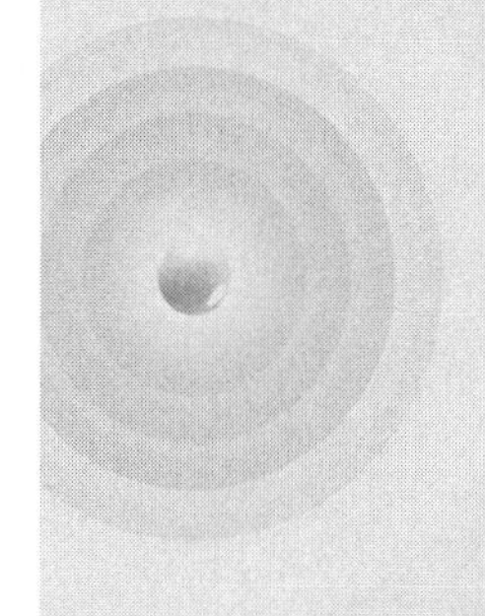

Questions & Answers

These questions are similar in style and content to those that you can expect in Unit Test 4. The number of questions is limited but they have been designed to test the majority of key facts and concepts covered in Unit 4.

Unit Test 4 is divided into sections A and B. Section A contains structured questions, with spaces at the end of each part of the question for your response. The number of questions in section A can vary, as can the mark for the individual questions. The total mark for Section A is 65.

Section B contains two or three longer questions, which are also divided into sections, worth a total of 35 marks. Each question is effectively a long structured question but with no spaces between each part for you to write your answer. Each part is worth more marks than those in section A.

The questions in the examination are likely to be set in an unfamiliar context, but do not be put off by this. The chemistry in each part of a question will relate to facts and concepts covered in this unit (or those covered previously at AS).

All the questions in this Question and Answer section are structured, without answer spaces. You should use the number of marks as a guide to the length of the answer required. If a question is worth 4 marks, then you should expect to write no more than four sentences, making four separate points. The marking scheme will then identify four key words or phrases that must be present for you to score the 4 marks. The question totals here can vary up to 36 marks. In Unit Test 4, it is unusual to find one question concentrating on one topic. However, the questions here could represent two or three separate structured questions from separate papers. The important point is that if you score well in each question, then at least you can be confident that you understand the topic.

Record your scores and convert them to a percentage and a grade. As a rough guide, you need to score 80% to achieve a grade A, 70% for a grade B and 60% for a grade C.

Examiner's comments

Grade-A answers are provided for all questions. Where appropriate, these are followed by examiner's comments (preceded by the icon e). These comments point out common errors, e.g. where a grade-C candidate is likely to lose marks, or suggest alternative answers that are acceptable.

Kinetics

(a) **The reaction of peroxodisulfate(VI) ions and iodide ions was investigated by carrying out three separate experiments at constant temperature.**

$$S_2O_8^{2-} + 2I^- \longrightarrow 2SO_4^{2-} + I_2$$

Experiment	Initial concentration of $S_2O_8^{2-}$/mol dm^{-3}	Initial concentration of I^-/mol dm^{-3}	Initial rate/ mol dm^{-3} s^{-1}
A	0.010	0.20	4.10×10^{-6}
B	0.020	0.20	8.20×10^{-6}
C	0.020	0.40	1.64×10^{-5}

(i) **Use the data to deduce the order with respect to the peroxodisulfate(VI) ions.** (2 marks)
(ii) **Use the data to deduce the order with respect to the iodide ions.** (2 marks)
(iii) **State the rate equation.** (1 mark)
(iv) **Calculate the rate constant and deduce its units.** (3 marks)

(b) **2-bromo-2-methylpropane reacts with sodium hydroxide according to the following equation:**

$$(CH_3)_3CBr + OH^- \longrightarrow (CH_3)_3COH + Br^-$$

The following data are the results of three experiments carried out at 25°C.

Experiment	Initial concentration of $(CH_3)_3CBr$/mol dm^{-3}	Initial concentration of OH^-/mol dm^{-3}	Initial rate of reaction/ mol dm^{-3} s^{-1}
1	0.010	0.20	3.00×10^{-3}
2	0.020	0.20	6.00×10^{-3}
3	0.040	0.40	1.20×10^{-2}

(i) **State the order with respect to 2-bromo-2-methylpropane.** (1 mark)
(ii) **State the order with respect to hydroxide ions.** (1 mark)
(iii) **State the rate equation.** (1 mark)
(iv) **Calculate the rate constant and deduce its units.** (3 marks)
(v) **Calculate the initial rate of the reaction when the initial concentration of $(CH_3)_3CBr$ is 0.035 mol dm^{-3} and the initial concentration of OH^- is 0.30 mol dm^{-3}.** (1 mark)
(vi) **State the effect, if any, on the rate constant of increasing the concentration of $(CH_3)_3CBr$ at a fixed temperature.** (1 mark)
(vii) **State, then explain, how a change in temperature affects the rate constant.** (4 marks)

(viii) The mechanism for the reaction consists of the two steps shown below:

Step 1 $(CH_3)_3CBr \longrightarrow (CH_3)_3C^+ + Br^-$

Step 2 $(CH_3)_3C^+ + OH^- \longrightarrow (CH_3)_3COH$

State which is the rate-determining step and explain your choice. (2 marks)

(c) The initial rate of reaction between substances A and B was measured in a series of experiments and the following rate equation was deduced:

rate = $k[A]^2[B]$

(i) Complete the table below for the reaction between A and B. (5 marks)

Experiment	Concentration of A/mol dm^{-3}	Concentration of B/mol dm^{-3}	Initial rate/ mol dm^{-3} s^{-1}
1	0.020	0.020	1.20×10^{-4}
2	0.040	0.040	
3	0.040		2.40×10^{-4}
4	0.030	0.060	
5		0.040	7.20×10^{-4}
6	0.080	0.080	

(ii) Use the data from experiment 1 to deduce a value for the rate constant, *k*, and state its units. (3 marks)

Total: 30 marks

Grade-A answer to Question 1

(a) (i) First order with respect to $S_2O_8^{2-}$ ✓. In experiments A and B, doubling $[S_2O_8^{2-}]$ and keeping $[I^-]$ constant leads to a doubling of the reaction rate ✓.

Since 2 marks are available, for the second mark you must indicate how you deduced the order. It is essential that you mention that the concentration of I^- remains constant.

(ii) First order with respect to I^- ✓. In experiments B and C, doubling $[I^-]$ and keeping $[S_2O_8^{2-}]$ constant leads to a doubling in the reaction rate ✓.

Similarly, as 2 marks are available, for the second mark you must indicate how you deduced the order. It is essential that you mention that the concentration of $S_2O_8^{2-}$ remains constant.

(iii) rate = $k[S_2O_8^{2-}][I^-]$ ✓

In the rate equation you must include the square brackets to indicate concentration in mol dm^{-3}.

(iv) $k = \dfrac{\text{rate}}{[S_2O_8^{2-}][I^-]} = \dfrac{4.10 \times 10^{-6}}{0.010 \times 0.20}$ ✓ $= 2.05 \times 10^{-3}$ ✓ mol^{-1} dm^3 s^{-1} ✓

The correct answer with the correct units scores 3 marks. If the answer is incorrect you can still score 1 mark for rearranging the rate equation and 1 mark for showing the correct units. The calculation shown here uses the data from experiment A. All of the experiments would give the same answer.

(b) (i) First order with respect to $(CH_3)_3CBr$ ✓

Only the statement 'first order' is required and only 1 mark is available. However, the reasoning behind the answer is as follows: look at experiments 1 and 2; $[(CH_3)_3CBr]$ is doubled and $[OH^-]$ is kept constant leads to a doubling of the reaction rate.

(ii) Zero order with respect to OH^- ✓

Only the statement 'zero order' is required and only 1 mark is available. However, the reasoning behind the answer is as follows: look at experiments 2 and 3; $[(CH_3)_3CBr]$ is doubled, which leads to a doubling of the reaction rate to 1.20×10^{-2}, so doubling OH^- at the same time must have had no effect on the reaction rate.

(iii) rate = $k[(CH_3)_3CBr]$ ✓

As OH^- is zero order, it does not appear in the rate equation.

(iv) $k = \frac{\text{rate}}{[(CH_3)_3CBr]} = \frac{3.00 \times 10^{-3}}{0.010}$ ✓ $= 0.30$ ✓ s^{-1} ✓

The calculation of the rate constant with the correct units scores 3 marks. If the answer is incorrect you can still score 1 mark for rearranging the rate equation and 1 mark for showing the correct units of the rate constant.

(v) rate = $k[(CH_3)_3CBr] = 0.30 \times 0.035 = 0.0105 = 1.05 \times 10^{-2}\,mol\,dm^{-3}\,s^{-1}$ ✓

If the answer to part **(iv)** was incorrect, this mark could still be obtained consequentially on the answer given in part **(iv)**.

(vi) Concentration has no effect on the rate constant ✓.

The key point to remember here is that only temperature alters the rate constant.

(vii) The rate constant increases with an increase in temperature ✓ because an increased temperature increases the average kinetic energy of the particles ✓, more particles now possess energy greater than the energy of activation ✓, leading to many more successful collisions ✓.

The increase in average kinetic energy of the particles would also lead to a greater collision frequency but this is a much less important factor than the increase in the number of particles which have an energy greater than the activation energy.

(viii) Step 1 ✓ because it contains the substance that appears in the rate equation ✓.

(c) (i)

Experiment	Concentration of A/mol dm^{-3}	Concentration of B/mol dm^{-3}	Initial rate/ mol dm^{-3} s^{-1}
1	0.020	0.020	1.20×10^{-4}
2	0.040	0.040	9.60×10^{-4} ✓
3	0.040	0.010 ✓	2.40×10^{-4}
4	0.030	0.060	8.10×10^{-4} ✓
5	0.035 ✓	0.040	7.20×10^{-4}
6	0.080	0.080	7.70×10^{-3} ✓

Completing a table of initial concentrations and initial rates is a common task, although it is likely that you would be asked to complete only one or two boxes rather than five. Marks are awarded for the final answers and no explanations are required. However, to aid your understanding, the answers, based on the rate equation, were achieved as follows:

- Experiment 2: the reaction is third-order overall, so doubling the concentration of both reactants between experiments 1 and 2 will lead to a 2^3 increase in reaction rate, that is, $8 \times 1.20 \times 10^{-4} = 9.60 \times 10^{-4}$.
- Experiment 3: comparing experiments 1 and 3, doubling the concentration of A should produce a quadrupling in the rate of reaction, that is, 4.80×10^{-4}. However, the rate is only half this value at 2.40×10^{-4}, so the concentration of B must have been halved.
- Experiment 4: comparing experiments 1 and 4, the concentration of A has increased by a factor of 1.5, so the rate will increase by $(1.5)^2 = 2.25$. The concentration of B has increased by a factor of 3, so the rate should increase overall by a factor of $2.25 \times 3 = 6.75$. This means the final answer is $6.75 \times 1.20 \times 10^{-4} = 8.10 \times 10^{-4}$.
- Experiment 5: comparing experiments 1 and 5, doubling the concentration of B should lead to a doubling in the reaction rate to 2.40×10^{-4}. The final rate is three times this value at 7.20×10^{-4}. This means that the concentration of A has increased by $\sqrt{3} = 1.732$. The concentration of A is $0.020 \times 1.732 = 0.03464 = 0.035$.
- Experiment 6: comparing experiments 2 and 6, the concentration of both reactants has doubled. The rate will therefore increase by a factor of 2^2 (= 4) for the increase in [A] and by a factor of 2 for the increase in [B], and so will increase by a factor of 8 overall. $8 \times 9.60 \times 10^{-4} = 7.68 \times 10^{-3}$.

(ii) $k = \dfrac{\text{rate}}{[A]^2[B]} = \dfrac{1.20 \times 10^{-4}}{0.020^2 \times 0.020}$ ✓ = 15 ✓ mol^{-2} dm^6 s^{-1} ✓

If the answer is incorrect then 1 mark will be awarded for rearranging the rate equation and 1 mark for the correct units. The data for all the experiments should produce the same rate constant.

Equilibria

(a) When 1.00 mol of ethanol and 1.00 mol of ethanoic acid were mixed together at 20°C, the reaction mixture at equilibrium was found to contain 0.667 mol of ethyl ethanoate. The total volume of the reaction mixture was $0.10\,dm^3$. The equation for the reaction is:

$$C_2H_5OH + CH_3COOH \rightleftharpoons CH_3COOC_2H_5 + H_2O$$

(i) Write an expression for the equilibrium constant, K_c, for this reaction. (1 mark)

(ii) Calculate the equilibrium concentrations, in $mol\,dm^{-3}$, of ethyl ethanoate and ethanol in the reaction mixture. (2 marks)

(iii) Calculate the value of K_c at this temperature. (2 marks)

(iv) In a separate experiment, 0.600 mol of ethanoic acid, 0.500 mol of ethanol, 0.60 mol of ethyl ethanoate, 0.400 mol of water and a small amount of concentrated sulfuric acid were mixed together at 20°C. At equilibrium, only 0.400 mol of ethanoic acid remained. The total volume of the reaction mixture was $0.10\,dm^3$. Calculate the value of K_c found in this experiment. (4 marks)

(b) Dinitrogen tetraoxide and nitrogen dioxide exist in the following equilibrium:

$$N_2O_4(g) \rightleftharpoons 2NO_2(g) \qquad \Delta H = +58\,kJ\,mol^{-1}$$

When 10.4 g of N_2O_4 was placed in a vessel of volume $4.50\,dm^3$ at a fixed temperature, 5.20 g of NO_2 was produced at equilibrium under a pressure of 100 kPa.

(i) Calculate:

- the number of moles of NO_2 at equilibrium
- the number of moles of N_2O_4 that reacted
- the original moles of N_2O_4
- the number of moles of N_2O_4 at equilibrium (4 marks)

(ii) Write an expression for the equilibrium constant, K_c, for this reaction. Calculate the value of K_c and state its units. (4 marks)

(iii) State the effect, if any, on both the yield of NO_2 and on the equilibrium constant, K_c, of increasing the temperature of the experiment. Give the reason for the effects you have stated. (4 marks)

(iv) State the effect, if any, on both the yield of NO_2 and on the equilibrium constant, K_c, of increasing the pressure. (2 marks)

Total: 23 marks

■ ■ ■

Grade-A answer to Question 2

(a) (i) $K_c = \dfrac{[CH_3COOC_2H_5][H_2O]}{[C_2H_5OH][CH_3COOH]}$ ✓

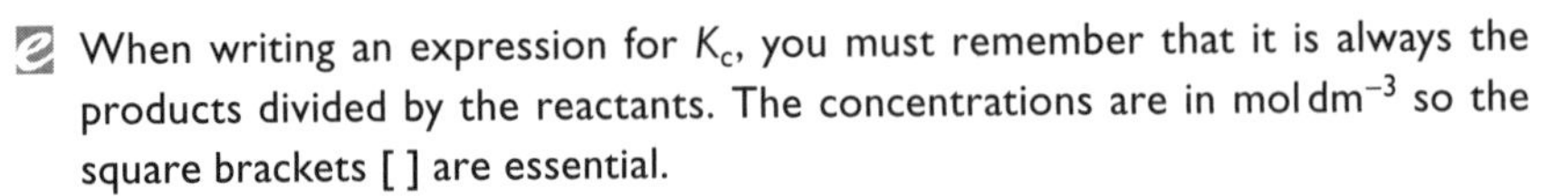

When writing an expression for K_c, you must remember that it is always the products divided by the reactants. The concentrations are in mol dm^{-3} so the square brackets [ ] are essential.

(ii) Concentration of ethyl ethanoate $= \dfrac{0.667}{0.10} = 6.67\ \text{mol dm}^{-3}$ ✓

Concentration of ethanol $= \dfrac{1 - 0.667}{0.10} = 3.33\ \text{mol dm}^{-3}$ ✓

When using K_c in calculations, it is essential that the concentrations are expressed in mol dm^{-3}. 0.667 mol in 0.10 dm^3 is a concentration of 6.67 mol dm^{-3}. In this reaction, ignoring the volume would not affect the final answer because the value of V (i.e. 0.10 dm^3) cancels out as there are equal numbers of particles on the two sides of the equation. However, in equilibria involving unequal numbers of reactant and product particles, the volume does not cancel out (see part b of this question).

(iii) $K_c = \dfrac{6.67 \times 6.67}{3.33 \times 3.33}$ ✓ = 4.0 ✓ (no units)

The equation shows that for every mole of $CH_3COOC_2H_5$ produced there will be the same number of moles of H_2O, so the top line of the K_c expression becomes 6.67 × 6.67. In a similar way, for every mole of ethanoic acid that reacts the same amount of ethanol reacts, so the bottom line of the K_c expression becomes 3.33 × 3.33. There are no units for K_c because there are the same number of particles on both sides of the equation.

(iv)

	C_2H_5OH +	CH_3COOH $\rightleftharpoons$	$CH_3COOC_2H_5$ +	H_2O
Initial moles	0.500	0.600	0.600	0.400
Equilibrium moles	$0.500 - x$	$0.600 - x$	$0.600 - x$	$0.400 + x$

If 0.400 mol of CH_3COOH remains, then the value of x must be 0.200 ✓.

Equilibrium moles	0.300	0.400	0.800	0.600 ✓

$K_c = \dfrac{(0.800/0.10) \times (0.600/0.10)}{(0.300/0.10) \times (0.400/0.10)}$ ✓ = 4.0 ✓

This is a more difficult calculation, but the principle is the same as above. Start with the original concentrations and determine how much has reacted (i.e. the value of x). This gives you how much you need to subtract from ethanol and how much you need to add to the original concentrations of ethyl ethanoate and water. The final answer of 4.0 scores 4 marks. However, if this is incorrect, then marks would be awarded for the working out, as indicated by the ticks in the grade-A answer. You would expect the value of K_c to be the same in parts (iii) and (iv) because the reactions were carried out at the same temperature.

(b) (i) Moles of NO_2 at equilibrium $= \dfrac{5.2}{46} = 0.113$ ✓

Moles of N_2O_4 reacted $= \dfrac{0.113}{2} = 0.0565$ ✓

Original moles of $N_2O_4 = \frac{10.4}{92} = 0.113$ ✓

Moles of N_2O_4 at equilibrium = 0.113 – 0.0565 = 0.0565 ✓

The key to this calculation is realising that when x amount of N_2O_4 breaks down it produces $2x$ amount of NO_2. So when 0.113 mol of NO_2 is produced, 0.113/2 (= 0.0565) mol of N_2O_4 has broken down.

(ii) $K_c = \frac{[NO_2]^2}{[N_2O_4]}$ ✓

$K_c = \frac{(0.113/4.5)^2}{0.0565/4.5}$ ✓ = 0.0502 = 5.02×10^{-2} ✓ mol dm^{-3} ✓

When calculating K_c, the most common mistake is to miss the total volume of the reaction vessel, which in this case is 4.5 dm^3. If you do not include this, your final answer will be 0.226 mol dm^{-3} and the only marks you could score here would be for the K_c expression and its units.

(iii) Increasing the temperature will increase the yield of NO_2 ✓ and increase the value of K_p ✓. The reaction is endothermic, so an increase in temperature causes the equilibrium to shift to the right-hand side ✓. The ratio of products to reactants increases so the value of K_c increases ✓.

Remember that concentration, temperature and pressure can all affect the yield of the product by shifting the equilibrium position. However, the only factor that alters K_c is temperature.

(iv) Increasing the pressure will decrease the yield of NO_2 ✓ but have no effect on K_c ✓.

In this case, an increase in pressure will cause the equilibrium to shift to the left-hand side because there are fewer particles on this side and this will reduce the pressure and oppose the change. Pressure does not alter K_c.

Acids and bases

(a) (i) Define the term Brønsted–Lowry acid. (1 mark)

(ii) Define pH. (1 mark)

(iii) Hydrochloric acid reacts with water according to the following equation:

$$HCl + H_2O \rightleftharpoons H_3O^+ + Cl^-$$

Write out the equation and label each species as an acid or a base. (1 mark)

(iv) Calculate the pH of a 0.500 mol dm^{-3} solution of hydrochloric acid. (1 mark)

(b) Apple juice has a pH of 3.52. It is assumed that apple juice contains a weak monoprotic acid, HA. A 25.0 cm^3 sample of apple juice was exactly neutralised by 28.5 cm^3 of 0.100 mol dm^{-3} sodium hydroxide using a suitable indicator.

(i) Define the following terms when applied to an acid:

- weak
- monoprotic (2 marks)

(ii) Assuming that apple juice contains this single monoprotic acid, HA, calculate the molar concentration of the acid in the juice. (3 marks)

(iii) Use the pH value to determine the molar concentration of hydrogen ions in apple juice. (2 marks)

(iv) Explain the difference between the two results for parts (ii) and (iii). (1 mark)

(v) Write an expression for K_a for the dissociation of the weak acid, HA. (1 mark)

(vi) Calculate a value for K_a and include its units. (3 marks)

(c) A 25.0 cm^3 sample of ethanoic acid (K_a = 1.76 × 10^{-5} mol dm^{-3}) is titrated with a 0.150 mol dm^{-3} solution of sodium hydroxide. The equivalence point occurs at pH 8 and the volume of sodium hydroxide added at this point is 20.45 cm^3.

(i) Explain the difference between the terms 'equivalence point' and 'end point'. (2 marks)

(ii) Use the information on the equivalence point to determine the original concentration of the ethanoic acid. (3 marks)

(iii) Calculate the pH of the acid before the addition of any sodium hydroxide. (3 marks)

(iv) Calculate the pH of the solution after the addition of 10.0 cm^3 of sodium hydroxide. (5 marks)

(v) Calculate the pH of the solution after the addition of 40.0 cm^3 of sodium hydroxide. (5 marks)

(d) (i) Explain what is meant by the term 'buffer solution'. (2 marks)

(ii) Give an example of a basic buffer. (2 marks)

(iii) Explain how an equimolar solution of nitrous acid (HNO_2) and sodium nitrite ($NaNO_2$) is able to maintain a constant pH on the addition of small amounts of sodium hydroxide. (3 marks)

(iv) Calculate the pH of a buffer solution made by mixing together 10.0 cm^3 of 1.00 mol dm^{-3} nitrous acid with 20.0 cm^3 of 2.00 mol dm^{-3} sodium nitrite solution. (K_a of nitrous acid = 4.57×10^{-4} mol dm^{-3}) (4 marks)

Total: 45 marks

■ ■ ■

Grade-A answer to Question 3

(a) (i) Proton donor ✓

The term 'H^+ donor' is an acceptable answer. A lone pair acceptor is incorrect, as this is the definition of a *Lewis* acid.

(ii) $pH = -\log_{10}[H^+(aq)]$ ✓

The minimum acceptable answer is '$pH = -\log[H^+]$'. The square brackets are absolutely essential in the definition because this indicates molar concentration of hydrogen ions, that is, mol dm^{-3}.

(iii) $HCl + H_2O \rightleftharpoons H_3O^+ + Cl^-$

acid   base   acid   base ✓

You have to identify all four species as either an acid or a base to obtain the mark.

(iv) $pH = -\log_{10}0.500 = 0.30$ ✓

(b) (i) Partially ionised in solution ✓; liberates one proton per molecule on dissociation ✓.

The terms 'partially' or 'slightly' and 'ionised' or 'dissociated' are acceptable. The term 'not fully' would be given no credit.

(ii) Moles of NaOH used $= \frac{MV}{1000} = \frac{0.100 \times 28.5}{1000} = 2.85 \times 10^{-3}$ ✓

Moles of acid HA in 25.0 cm^3 = 2.85×10^{-3} ✓

Moles of HA in 1 dm^3, i.e. $[HA] = 2.85 \times 10^{-3} \times \frac{1000}{25} = 0.114$ mol dm^{-3} ✓

The correct answer of 0.114 mol dm^{-3} scores 3 marks. However, always show your working, because marks can be gained for each stage of the calculation. There are three definite steps in this calculation: (1) work out moles of alkali used; (2) use the 1:1 ratio in the equation; and (3) scale up the number of moles to molar concentration.

(iii) $pH = -\log_{10}[H^+(aq)]$, so $[H^+(aq)] = 10^{-pH}$ ✓

$[H^+(aq)] = 10^{-3.52} = 3.02 \times 10^{-4}$ mol dm^{-3} ✓

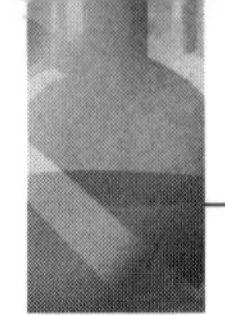

(iv) The acid is weak. It is only partially ionised, so it liberates few H^+ ions in solution ✓.

The original concentration is $0.114\,mol\,dm^{-3}$ and the number of H^+ ions in solution is $3.02 \times 10^{-4}\,mol\,dm^{-3}$, so only about 0.3% of the acid has dissociated, making it a weak acid. However, all the H^+ ions react with NaOH.

(v) $K_a = \frac{[H^+][A^-]}{[HA]}$ ✓

Common mistakes include using round brackets instead of square brackets and trying to simplify the expression to

$$K_a = \frac{[H^+]^2}{[HA]}$$

The latter may be used in calculations but not as the expression for K_a.

(vi) $K_a = \frac{[H^+]^2}{[HA]} = \frac{(3.02 \times 10^{-4})^2}{0.114}$ ✓ $= 8.00 \times 10^{-7}$ ✓ $mol\,dm^{-3}$ ✓

In this calculation it is assumed that the equilibrium concentration of H^+ is equal to the equilibrium concentration of A^-, that is, $[H^+] = [A^-]$ (so the top line of the equation now becomes $[H^+]^2$). It is also assumed that the amount of dissociation of HA is so small that the original concentration of HA is equal to the equilibrium concentration of HA, that is, $[HA]_{orig} = [HA]_{eqm}$. In calculations of K_a, remember to include the units for K_a at the end.

(c) (i) The equivalence point occurs when stoichiometric amounts (equal numbers of moles) of acid and base are mixed together ✓. The end point occurs when the indicator changes colour ✓.

Any expression that implies that equal numbers of moles of acid and alkali have been mixed is acceptable for the definition of the equivalence point. Indicators are chosen so that the colour change (end point) coincides with the equivalence point.

(ii) Moles of NaOH added $= \frac{MV}{1000} = \frac{0.15 \times 20.45}{1000} = 3.0675 \times 10^{-3}$ ✓

Moles of CH_3COOH in $25.0\,cm^3 = 3.0675 \times 10^{-3}$ ✓

$[CH_3COOH] = 3.0675 \times 10^{-3} \times \frac{1000}{25.0} = 0.1227 = 0.123\,mol\,dm^{-3}$ ✓

The correct answer scores 3 marks. Remember your working, because if your final answer is incorrect then credit will be given for showing any of the three steps: moles of alkali added; using the 1:1 ratio in the equation; and scaling up the moles of acid to molar concentration.

(iii) $K_a = \frac{[CH_3COO^-][H^+]}{[CH_3COOH]} = \frac{[H^+]^2}{[CH_3COOH]}$ ✓

$[H^+] = \sqrt{K_a \times [CH_3COOH]} = \sqrt{1.76 \times 10^{-5} \times 0.1227} = 1.469 \times 10^{-3}$ ✓

$pH = -\log_{10}(1.469 \times 10^{-3}) = 2.83$ ✓

The correct answer scores 3 marks. This calculation depends on the assumption that the number of moles of H^+ equals the number of moles of CH_3COO^- (so the K_a expression is simplified to include $[H^+]^2$) and that the original concentration of CH_3COOH is the same as the concentration of CH_3COOH at equilibrium (because only a tiny amount of the CH_3COOH has dissociated). This means the expression $[H^+] = K_a \times \sqrt{[CH_3COOH]}$ can be used to calculate pH. A common answer is **5.67** (failing to take the square root) and this scores only 2 marks.

(iv) Original moles of $CH_3COOH = \frac{MV}{1000} = \frac{0.1227 \times 25.0}{1000} = 3.0675 \times 10^{-3}$

Moles of NaOH added $= \frac{0.15 \times 10.0}{1000} = 1.50 \times 10^{-3}$ ✓

Moles of CH_3COO^- formed $= 1.50 \times 10^{-3}$ ✓

Moles of CH_3COOH remaining $= (3.0675 \times 10^{-3}) - (1.50 \times 10^{-3}) = 1.5675 \times 10^{-3}$ ✓

$K_a = \frac{[CH_3COO^-][H^+]}{[CH_3COOH]}$, so $[H^+] = K_a \times \frac{[CH_3COOH]}{[CH_3COO^-]}$

$[H^+] = 1.76 \times 10^{-5} \times \frac{1.5675 \times 10^{-3}}{1.50 \times 10^{-3}} = 1.8392 \times 10^{-5}$ ✓

$pH = -\log_{10}(1.8392 \times 10^{-5}) = 4.7353 = 4.74$ ✓

This is probably one of the most difficult calculations at A2 and only the most able candidates will score full marks on this section. The correct answer of **4.74** scores 5 marks but if the final answer is incorrect then credit will be given for the working.

(v) Moles of $CH_3COOH = 3.0675 \times 10^{-3}$

Moles of NaOH added $= \frac{0.15 \times 40.0}{1000} = 6.00 \times 10^{-3}$ ✓

Excess moles of NaOH $= (6.00 \times 10^{-3}) - (3.0675 \times 10^{-3}) = 2.9325 \times 10^{-3}$ ✓

Total volume of the solution $= 25.0 + 40.0 = 65.0\,cm^3$

$[OH^-] = 2.9325 \times 10^{-3} \times \frac{1000}{65.0} = 0.0451\,mol\,dm^{-3}$ ✓

$[H^+] = \frac{K_W}{[OH^-]} = \frac{1.0 \times 10^{-14}}{0.0451} = 2.217 \times 10^{-13}$ ✓

$pH = -\log_{10}(2.217 \times 10^{-13}) = 12.65 = 12.7$ ✓

This is another difficult calculation. Again, the correct answer scores 5 marks. There are definite steps to remember in calculations of this type: moles of acid added; moles of alkali added; excess alkali (or sometimes excess acid); total volume of solution formed; molar concentration of OH^-; using K_w to determine the molar concentration of H^+; and using $-\log_{10}[H^+]$ to determine pH. The most common mistake is failing to scale up OH^- to molar concentration. If you fail to do this then you will get **$[H^+] = 3.41 \times 10^{-12}$** and **pH = 11.47**. This is an incorrect answer but you will still be awarded 3 out of a possible 5 marks.

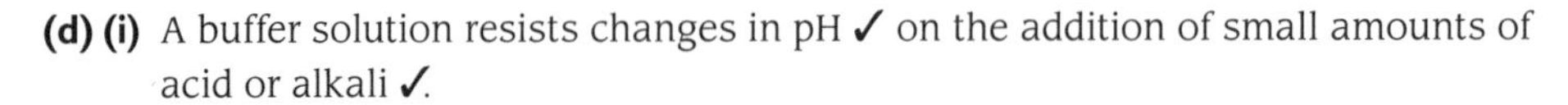

(d) (i) A buffer solution resists changes in pH ✓ on the addition of small amounts of acid or alkali ✓.

The two key points in this definition are 'resisting pH change' and 'adding small amounts'.

(ii) Ammonia ✓ and ammonium chloride ✓

Any weak alkali and its salt can be used as an example of a basic buffer, another example would be methylamine and methylammonium chloride.

(iii) The following equilibrium exists between nitrous acid and sodium nitrite:

$$HNO_2 \rightleftharpoons H^+ + NO_2^-$$ ✓

The OH^- ions that are added react with the H^+ ions ✓. However, the pH does not change because some HNO_2 dissociates and the equilibrium position shifts to the right ✓ to replace the H^+ ions.

The key point is that the buffer solution contains a large amount of undissociated acid (HNO_2) and a large amount of the anion NO_2^- (which has come from the salt). Addition of OH^- removes H^+ but the equilibrium position moves to the right to replace the H^+. If H^+ is added it reacts with the NO_2^- and the equilibrium position shifts to the left-hand side and removes the added H^+. In both cases, the pH change has been resisted.

(iv) $K_a = \dfrac{[H^+][NO_2^-]}{[HNO_2]}$ ✓

$$\text{Moles } HNO_2 = \frac{1.00 \times 10.0}{1000} = 0.01$$

$$\text{Moles } NaNO_2 = \frac{2.00 \times 20.0}{1000} = 0.04$$

$$\therefore K_a = \frac{[H^+] \times [0.04]}{[0.01]} = 4.57 \times 10^{-4}$$ ✓

$$[H^+] = \frac{4.57 \times 10^{-4}}{4} = 1.1425 \times 10^{-4}$$ ✓; $pH = -\log_{10}(1.1425 \times 10^{-4}) = 3.94$ ✓

An alternative method would be as follows:

$K_a = 4.57 \times 10^{-4}$, so $pK_a = 3.34$ ✓

$pH = pK_a + \log_{10}\dfrac{[\text{salt}]}{[\text{acid}]}$ ✓ $= 3.34 + \log_{10}\dfrac{[0.04]}{[\text{acid}]}$ ✓ $= 3.34 + \log 4 = 3.94$ ✓

Nomenclature and isomerism in organic chemistry

(a) Name the following compounds.

(i) $H_3C—C(OH)(H)—C \equiv N$

(ii) $H_3C—CH(CH_3)—C(=O)—H_2N$

(iii) $H_3C—CH(NH_2)—C(=O)—HO$

(iv) $H_3C—CH_2—CH_2—NH_2$

(v) $H_3C—CH_2—CH_2—C(=O)—Cl$

(vi) $H_3C—C_6H_4—CH_3$

(6 marks)

(b) Explain the terms 'structural isomerism' and 'stereoisomerism'. (4 marks)

(c) (i) Draw and name the four isomeric esters of molecular formula $C_4H_8O_2$. (8 marks)

(ii) Draw and name the two carboxylic acids that also have the molecular formula $C_4H_8O_2$. (4 marks)

(d) Draw all the structural isomers of C_4H_9Br. State the name of the isomer that is capable of exhibiting optical isomerism, draw the two enantiomers of this compound and explain how these two isomers could be distinguished from each other. (10 marks)

(e) (i) Draw the two *E–Z* isomers of but-2-enal and label them *E* or *Z*. (4 marks)

(ii) Draw the two structural isomers of but-2-enal that do not exhibit *E–Z* isomerism but still contain an alkene and an aldehyde functional group. Explain why these isomers do not exhibit *E–Z* isomerism. (3 marks)

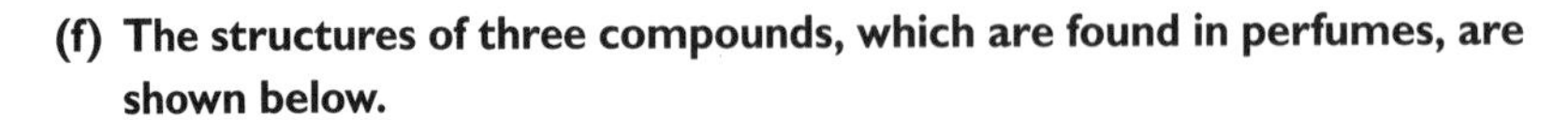

(f) The structures of three compounds, which are found in perfumes, are shown below.

Compound A — Citronellol

Compound B — Linalool

Compound C — Geraniol

Use the letters A, B and C to answer the questions below.

(i) Which of these compounds are structural isomers? Give the molecular formula of these isomers.

(ii) Which of these compounds will exhibit *E–Z* isomerism?

(iii) Which of these compounds will exhibit optical isomerism? (5 marks)

Total: 44 marks

Grade-A answer to Question 4

(a) (i) 2-hydroxypropanenitrile ✓

(ii) 2-methylpropanamide ✓

(iii) 2-aminopropanoic acid ✓

(iv) Propylamine ✓

(v) Butanoyl chloride ✓

(vi) 1,4-dimethylbenzene ✓

You would not be asked to name as many organic compounds as this in an examination, but it is essential that you are able to correctly relate any names to formulae that appear in a question.

(b) Structural isomerism occurs when molecules have the same molecular formula ✓ but different structural formulae ✓. Stereoisomerism occurs when molecules have the same molecular formula and the same structural formula ✓ but a different spatial arrangement of their bonds ✓.

You may not be asked to explain both these types of isomerism in an examination, but it is essential that you understand both terms and can recognise the different types.

(c) (i)

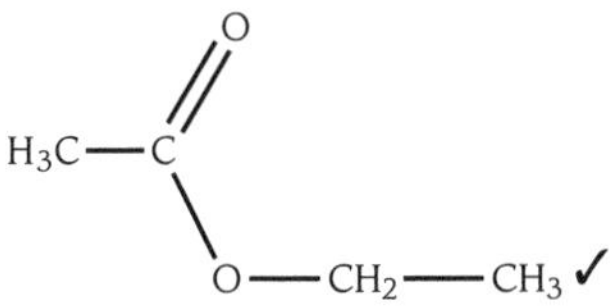

$H_3C-C(=O)-O-CH_2-CH_3$ ✓

Ethyl ethanoate ✓

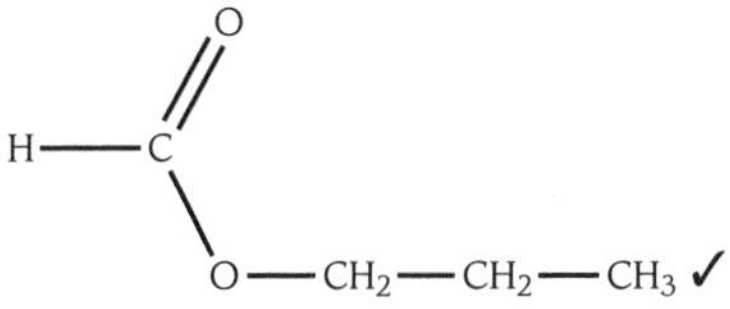

$H-C(=O)-O-CH_2-CH_2-CH_3$ ✓

Propyl methanoate ✓

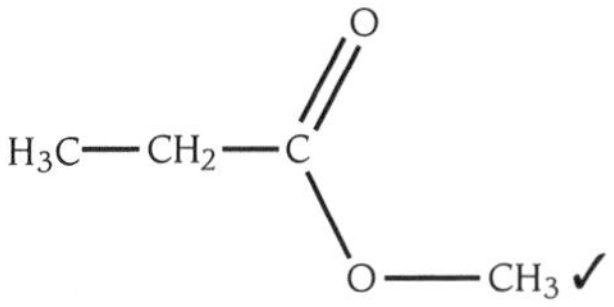

$H_3C-CH_2-C(=O)-O-CH_3$ ✓

Methyl propanoate ✓

$H-C(=O)-O-CH(CH_3)-CH_3$ ✓

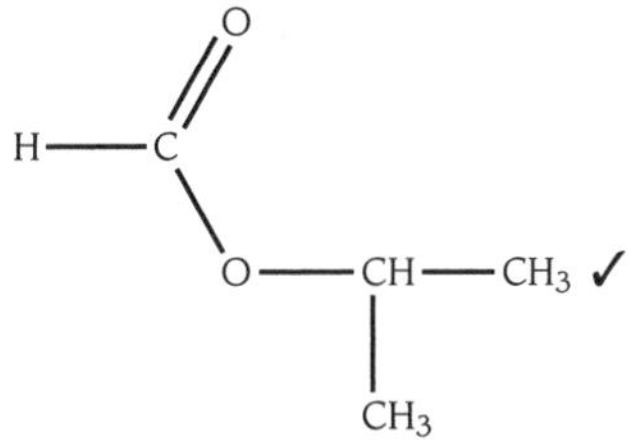

Methylethyl methanoate ✓

When drawing esters, remember to start with the functional group

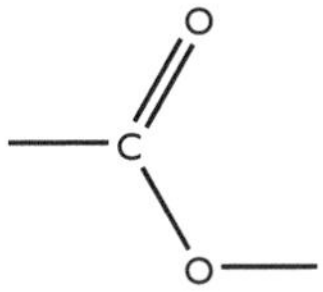

$-C(=O)-O-$

then place the remaining carbons each side of this group. You cannot attach hydrogen to the oxygen because this would produce a carboxylic acid. The most difficult ester to identify and name is the branched ester. Only the most able candidates will score this mark.

(ii)

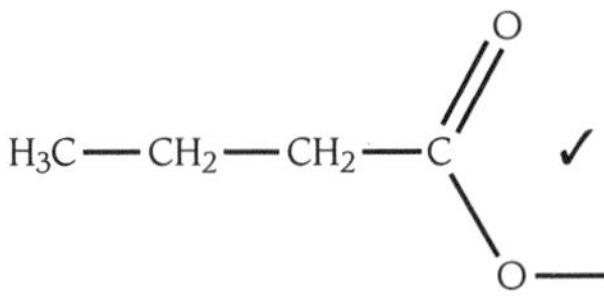

$H_3C-CH_2-CH_2-C(=O)-O-H$ ✓

Butanoic acid ✓

$H_3C-CH(CH_3)-C(=O)-O-H$ ✓

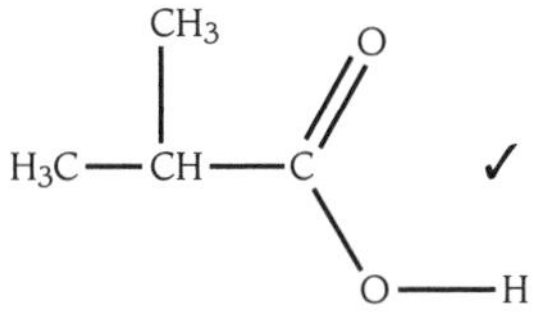

Methylpropanoic acid ✓

Always draw the straight-chain carboxylic acid with the –COOH at the end of the chain, then consider if branching is possible in the hydrocarbon chain. Remember that the carbon of the carboxylic acid group is position 1.

(d) $H_3C-CH_2-CH_2-CH_2-Br$ ✓

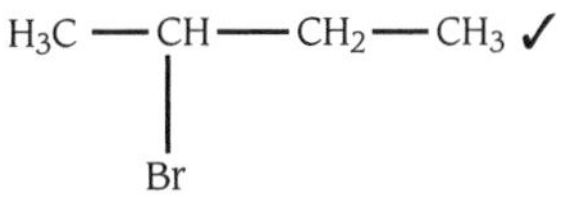

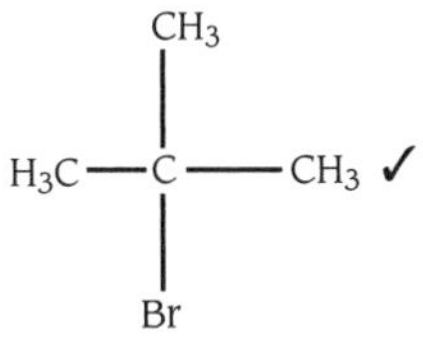

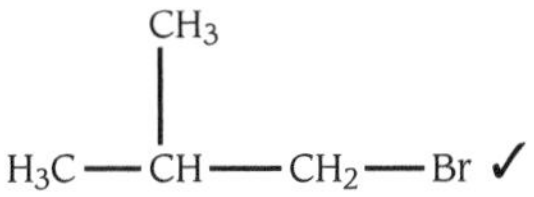

$H_3C-CH(Br)-CH_2-CH_3$ ✓

$H_3C-C(CH_3)(Br)-CH_3$ ✓

$H_3C-CH(CH_3)-CH_2-Br$ ✓

The compound capable of exhibiting optical isomerism is 2-bromobutane ✓. The two enantiomers of 2-bromobutane are

Mirror

H | C | H_3C | Br | CH_2—CH_3 ✓ ⋮ H_3C—CH_2 | C | H | Br | CH_3 ✓

The enantiomers can be distinguished by their ability to rotate ✓ the plane of plane-polarised light ✓ in opposite directions ✓.

You will not be awarded full marks if you draw more than four structural isomers: four correct isomers and one extra isomer will score 3 marks, four correct isomers and two extra isomers will score 2 marks, and so on.

Only 2-bromobutane contains a carbon atom with four different groups attached. When drawing the enantiomers you must emphasise the importance of the asymmetric carbon atom and show the four different groups attached in a tetrahedral fashion. Enantiomers have the same chemical and physical properties but they differ in their effect on plane-polarised light. Using the word 'reflect' rather than 'rotate' is a common mistake.

(e) (i) CH_3, H on C=C; H, C(H)=O on C=C ✓ *E* ✓

CH_3, H on C=C; C(H)=O, H on C=C ✓ *Z* ✓

In order to exhibit *E–Z* isomerism, the carbon atoms attached to the double bond must have different groups attached.

(ii) H, H on C=C; C(H)=O, CH_3 on C=C ✓

H, H on C=C; CH_2—C(H)=O, H on C=C ✓

These structural isomers do not show *E–Z* isomerism as they have two hydrogen atoms attached to one of the carbon atoms of the double bond ✓.

(f) (i) Compounds A and C ✓, molecular formula is $C_{10}H_{20}O$ ✓

You have been asked to apply the basic principles to three unusual compounds that you will not have met before. Compounds A and C are structural isomers because they have the same molecular formula but different structural formulae. Compound B has a molecular formula of $C_{10}H_{18}O$.

(ii) Compound C only ✓

e Compound C has four different groups attached to the carbon atoms involved in the carbon–carbon double bond and will exhibit *E–Z* isomerism. Compounds A and B will not exhibit geometric isomerism. Compound A has two methyl groups attached to a carbon atom involved in the carbon–carbon double bond. Compound B has two methyl groups attached to a carbon atom involved in one carbon–carbon double bond and two hydrogen atoms attached to a carbon atom in the other double bond.

(iii) Compounds A ✓ and B ✓

e Compounds A and B both possess asymmetric carbon atoms, that is, a carbon atom with four different groups attached, but compound C possesses no such carbon atom.

Aldehydes and ketones

(a) Consider the carbonyl compounds labelled A to C below.

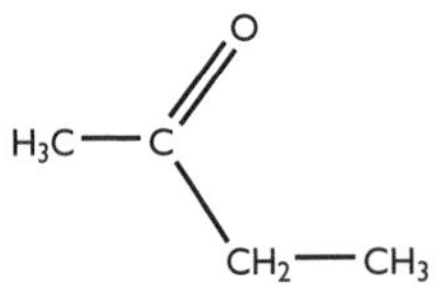

Compound A 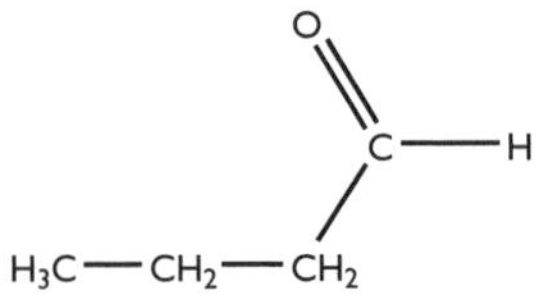 | Compound B 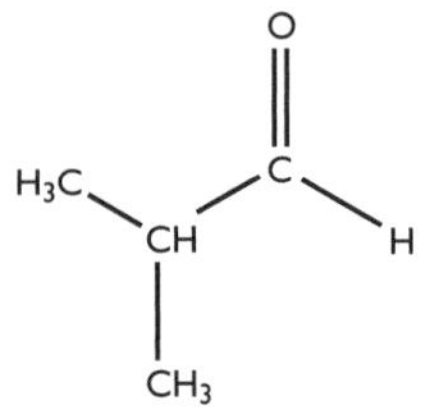 | Compound C

(i) **Name compounds A, B and C.** (3 marks)

(ii) **Write equations for the reaction of compound B with acidified $K_2Cr_2O_7$ and compound C with $NaBH_4$. Show the structure of the final product clearly.** (4 marks)

(iii) **Write an equation for the oxidation of the appropriate alcohol to produce compound A.** (2 marks)

(iv) **State which of the compounds labelled A to C would react with ammoniacal silver nitrate (Tollens' reagent). Explain why they can react and state the observations for a positive result.** (4 marks)

(v) **What colour change would be observed if compound C were warmed with Fehling's solution?** (1 mark)

(vi) **State the name of, then draw, the mechanism for the reaction of compound B with HCN.** (4 marks)

(b) Compound X has the following structure:

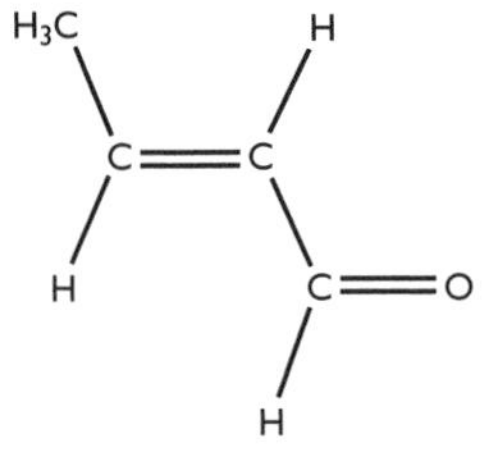

(i) **Name compound X.** (1 mark)

(ii) **Write equations for the reactions of compound X with Br_2 and with HCN.** (2 marks)

(iii) **Give simple structural formulae for the products formed when compound X is:**
- **reduced with $NaBH_4$**
- **oxidised with acidified $K_2Cr_2O_7$** (2 marks)

(iv) **Draw the mechanism for the reaction of compound X with $NaBH_4$.** (3 marks)

Total: 26 marks

Grade-A answer to Question 5

(a) (i) A = butanone ✓; B = butanal ✓; C = methylpropanal ✓

There is no need to include the numbers to indicate the position of the functional group. Compound A is a ketone and the C=O group can only occur at the 2 position. Compounds B and C are aldehydes and the CHO group is always numbered 1.

(ii)

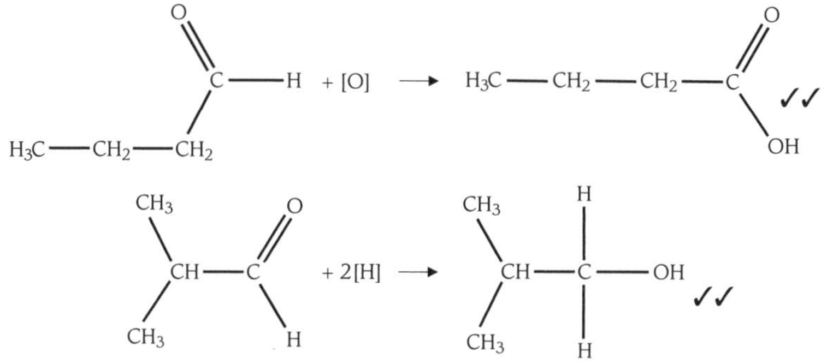

$$H_3C{-}CH_2{-}CH_2{-}C(=O){-}H + [O] \longrightarrow H_3C{-}CH_2{-}CH_2{-}C(=O){-}OH \quad ✓✓$$

$$(CH_3)_2CH{-}C(=O){-}H + 2[H] \longrightarrow (CH_3)_2CH{-}CH_2{-}OH \quad ✓✓$$

In each case there is 1 mark for the product and 1 mark for balancing the equation with the appropriate symbol. When compound B is oxidised, [O] is used for the oxidising agent and when compound C is reduced, [H] is used for the reducing agent. Do not try to include $K_2Cr_2O_7$ and $NaBH_4$ in your equations.

(iii)

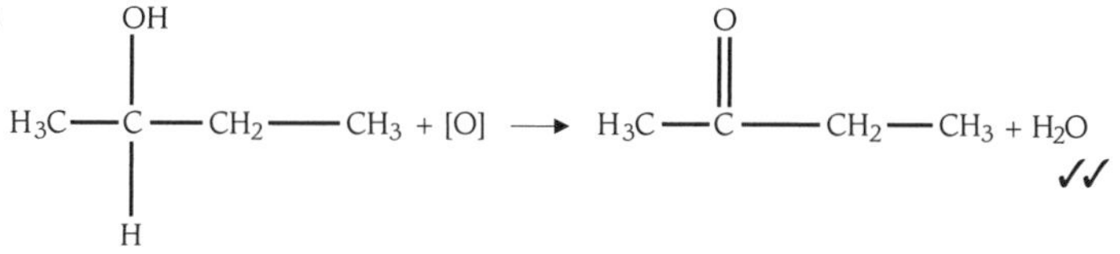

$$H_3C{-}CH(OH){-}CH_2{-}CH_3 + [O] \longrightarrow H_3C{-}C(=O){-}CH_2{-}CH_3 + H_2O \quad ✓✓$$

There is 1 mark for the structure of the alcohol and 1 mark for the equation. Again, use [O] for the oxidising agent. The most common mistake is the omission of H_2O from the balanced equation.

(iv) B ✓ and C ✓. Aldehydes can be further oxidised to a carboxylic acid ✓. A silver mirror is produced ✓.

The ability of aldehydes to be readily oxidised by mild oxidising agents provides the main method of distinguishing between aldehydes and ketones.

(v) Blue solution to red precipitate ✓

Fehling's solution is the alternative test, distinguishing between aldehydes and ketones. The red precipitate produced is copper(I) oxide.

(vi) Nucleophilic addition ✓

$CH_3CH_2CH_2-C(=O)H + :\bar{C}N \longrightarrow CH_3CH_2CH_2-C(\bar{O}:)(CN)-H \xrightarrow{H^+} CH_3CH_2CH_2-C(OH)(CN)-H$ ✓ ✓ ✓

The mechanism must be drawn carefully. The 3 marks are for the curly arrows shown, but these will not be awarded if the lone pairs on the cyanide ion and on the oxygen in the intermediate structure are not shown.

(b) (i) But-2-enal ✓

But-2-enal is an acceptable answer. However, if you name it as the geometric isomer then it *must* be named as *E*-but-2-enal.

(ii) $CH_3CH{=}CHCHO + Br_2 \longrightarrow CH_3CHBrCHBrCHO$ ✓
$CH_3CH{=}CHCHO + HCN \longrightarrow CH_3CH{=}CHCH(OH)CN$ ✓

It is quite common to see questions on the chemistry of compounds with more than one functional group. You must treat these functional groups separately. In the first reaction it is an alkene reacting with bromine and the rest of the molecule remains unchanged. In the second reaction it is an aldehyde reacting with HCN and the alkene double bond is unchanged.

(iii) $CH_3CH{=}CHCH_2OH$ ✓
$CH_3CH{=}CHCOOH$ ✓

Reduction products depend on the reagent used. $NaBH_4$ only attacks carbonyl compounds. When trying to predict the oxidation products using $K_2Cr_2O_7$, treat but-2-enal like any other aldehyde. Ignore the rest of the molecule, because it will remain unchanged.

(iv)

$CH_3CH{=}CH-C(=O)H + :\bar{H} \longrightarrow CH_3CH{=}CH-C(\bar{O}:)(H)-H \xrightarrow{H^+} CH_3CH{=}CH-C(OH)(H)-H$ ✓ ✓ ✓

Be careful when drawing mechanisms. The 3 marks will be awarded for the curly arrows provided that the lone pairs are included on the H^- ion and on the oxygen atom in the intermediate structure.

Carboxylic acids and esters

(a) (i) Draw two isomeric esters with the molecular formula $C_3H_6O_2$. (2 marks)

(ii) For each ester, write an equation for its preparation from suitable reagents. State the catalyst needed for both reactions. (5 marks)

(iii) For either one of these esters, predict the products of its alkaline hydrolysis. (2 marks)

(iv) Write equations for the reaction of the carboxylic acid CH_3CH_2COOH with:

- **sodium hydroxide**
- **sodium carbonate**
- **propan-2-ol** (3 marks)

(b) The structure of methylethyl methanoate is shown below:

(i) Deduce the molecular and empirical formulae of the ester. (2 marks)

(ii) Draw the structure of propyl methanoate, a structural isomer of methylethyl methanoate. State how many peaks you would expect in the ^{13}C NMR spectra of these two isomers and explain why they do not show the same number of peaks. (4 marks)

(c) 2-hydroxypropanoic acid, obtained from sour milk, has the ability to rotate plane-polarised light.

(i) Draw the structural formula of this acid and explain why it has this property. (3 marks)

(ii) Predict the structure of the compounds formed when 2-hydroxypropanoic acid reacts, in the presence of a strong acid catalyst, separately with:

- **ethanol**
- **ethanoic acid** (2 marks)

(d) Animal fats and vegetable oils are esters of long-chain carboxylic acids and can be used to produce soaps and biodiesel. For *each* of these products:

(i) give the reagents and conditions necessary to obtain them

(ii) give the name of the other product of the reactions

(iii) name the type of chemical process involved (7 marks)

Total: 30 marks

Grade-A answer to Question 6

(a) (i)

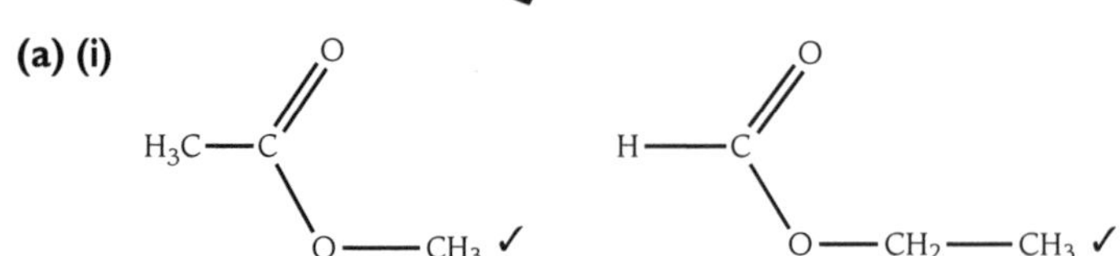

There are only two possible answers.

(ii) $CH_3COOH + CH_3OH \longrightarrow CH_3COOCH_3 + H_2O$ ✓ ✓
$HCOOH + CH_3CH_2OH \longrightarrow HCOOCH_2CH_3 + H_2O$ ✓ ✓
Concentrated sulfuric acid is the catalyst ✓.

The principle to apply here is that an alcohol reacts with an acid to produce an ester and water.

(iii) $CH_3COO^-Na^+$ ✓ and CH_3OH ✓ *or* $HCOO^-Na^+$ ✓ and C_2H_5OH ✓

The basic principle is that when an ester undergoes alkaline hydrolysis the bond breaks as shown below:

This part forms the salt R^1COO^-

This part forms the alcohol R^2OH

C–O bond broken

(iv) $CH_3CH_2COOH + NaOH \longrightarrow CH_3CH_2COO^-Na^+ + H_2O$ ✓
$2CH_3CH_2COOH + Na_2CO_3 \longrightarrow 2CH_3CH_2COO^-Na^+ + CO_2 + H_2O$ ✓
$CH_3CH_2COOH + CH_3CH(OH)CH_3 \longrightarrow CH_3CH_2COOCH(CH_3)_2 + H_2O$ ✓

Propanoic acid behaves as a typical acid, forming a salt and water with sodium hydroxide and liberating carbon dioxide from sodium carbonate solution. In the final reaction it is reacting with an alcohol to form an ester and water. The final product is a branched ester because the alcohol attaches to the carboxylic acid using the oxygen atom. The branched ester must be shown in the equation to gain the final mark.

(b) (i) The molecular formula is $C_4H_8O_2$ ✓ and the empirical formula is C_2H_4O ✓.

The molecular formula is the actual number of atoms of each element in the compound and the empirical formula is the simplest possible ratio of the atoms of each element in the compound.

(ii)

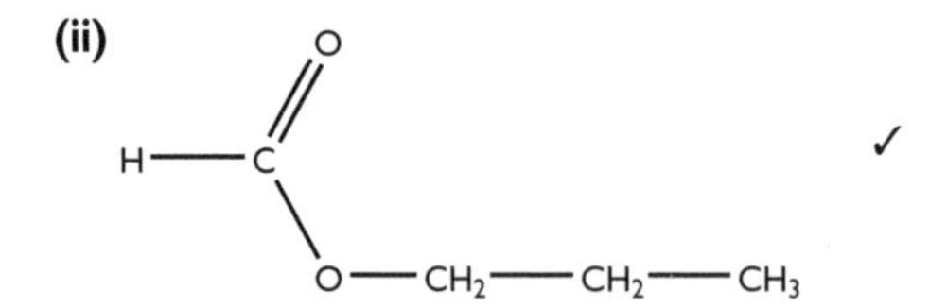

✓

Propyl methanoate has 4 peaks ✓, methylethyl methanoate has 3 peaks ✓. Both have four carbon atoms but in methylethyl methanoate those in the two methyl groups are equivalent ✓.

(c) (i)

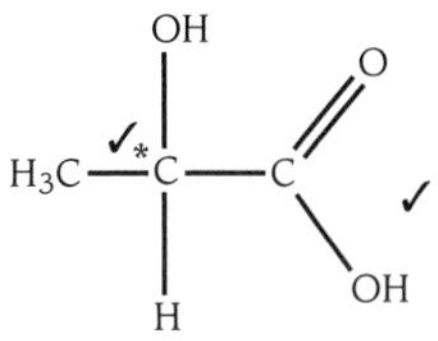

H_3C—C*(OH)(H)—C(=O)OH ✓ ✓

2-hydroxypropanoic acid is optically active because it is a chiral molecule ✓, that is, the carbon shown with an asterisk is asymmetric because it has four different groups attached.

To be optically active, a molecule must possess a chiral centre, that is, an asymmetric carbon atom.

(ii) With ethanol:

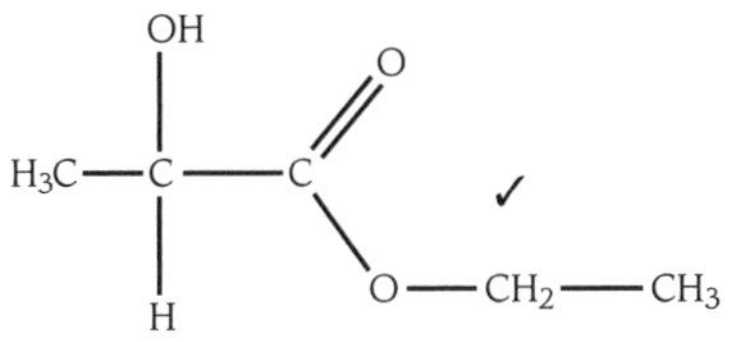

H_3C—C(OH)(H)—C(=O)—O—CH_2—CH_3 ✓

With ethanoic acid:

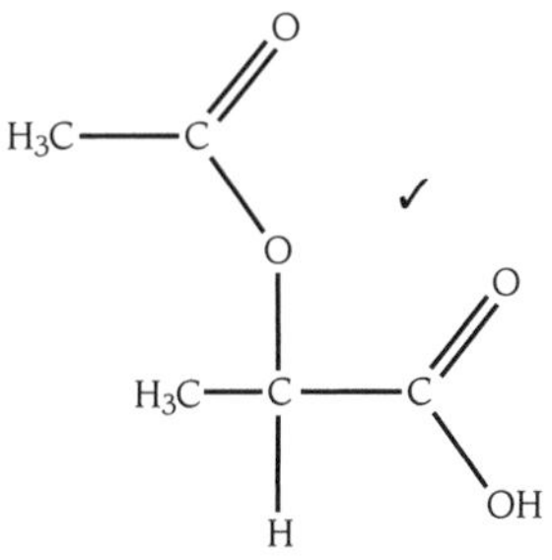

H_3C—C(=O)—O—C(H_3C)(H)—C(=O)OH ✓

The compound 2-hydroxypropanoic acid contains an –OH group and a –COOH group. Each group can form an ester, depending on the reagent added. The –COOH group will react with ethanol to form an ester and the –OH group will react with ethanoic acid to form an ester.

(d) (i) Soaps: sodium hydroxide ✓; biodiesel: catalyst/excess ✓ methanol ✓. Both reactions need heat ✓.

(ii) Propane-1,2,3-triol ✓

(iii) Soaps: alkaline hydrolysis ✓; biodiesel: transesterification ✓.

Both of these reactions are important industrial processes. Make sure that you can recall them both and take care not to get them confused.

Acylation

(a) (i) **Write an equation for the formation of ethyl ethanoate from ethanoyl chloride and ethanol.** (1 mark)

(ii) **Name and outline a mechanism for the reaction that takes place in part (i).** (5 marks)

(iii) **Suggest why the reaction in part (i) is a more efficient way of preparing ethyl ethanoate than the reaction between ethanoic acid and ethanol.** (1 mark)

(b) (i) **Write an equation for the reaction of propanoyl chloride with ethylamine.** (2 marks)

(ii) **Outline a mechanism for the reaction that takes place in part (i).** (4 marks)

(iii) **Suggest the structures of the acyl chlorides and amines used to produce the following amide derivatives X and Y:**

X: $H_3C—CH_2—CH_2—C(=O)—N(H)—C_6H_5$

Y: $C_6H_5—C(=O)—N(H)—CH(CH_3)_2$

(4 marks)

(c) **The compound 2-hydroxybenzenecarboxylic acid can be converted into two products, aspirin and oil of wintergreen, which are used as medicines.**

Aspirin

Oil of wintergreen

(i) **Write equations for the conversion of 2-hydroxybenzenecarboxylic acid into aspirin and into oil of wintergreen.** (4 marks)

(ii) **State the name of the catalyst used in the production of oil of wintergreen.** (1 mark)

(iii) **State the name of an alternative reagent for the production of aspirin.** (1 mark)

(d) **Consider the compounds A to C.**

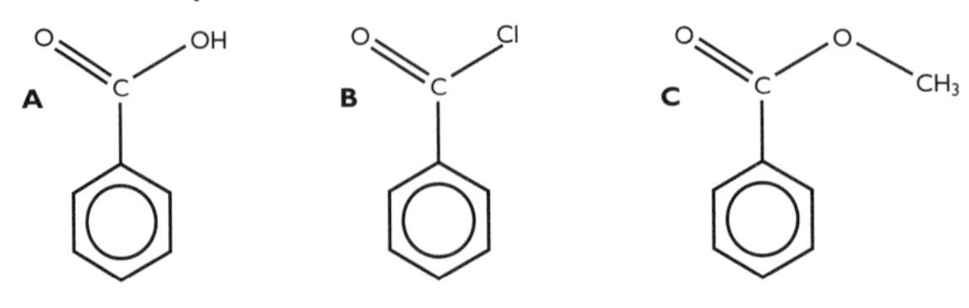

State the letter of the compound that will:

- **react with hot sodium hydroxide to produce methanol as one of the products**
- **react with ammonia to produce an amide**
- **react with cold water to produce compound A**

(3 marks)

Total: 26 marks

Grade-A answer to Question 7

(a) (i) $CH_3COCl + C_2H_5OH \longrightarrow CH_3COOC_2H_5 + HCl$ ✓

Acyl chlorides and alcohols react to produce an ester. The most common mistake is the omission of HCl from the equation.

(ii) Nucleophilic addition–elimination ✓

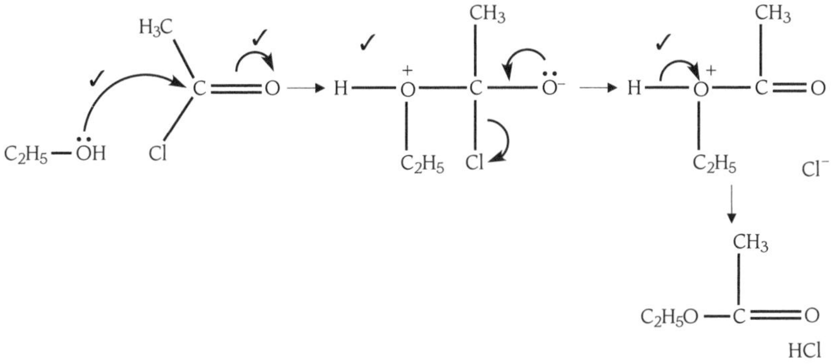

C_2H_5—ÖH ; H_3C, Cl, C=O → H—$\overset{+}{O}$—C—Ö$^-$ (with CH_3, C_2H_5, Cl) → H—$\overset{+}{O}$—C=O (with CH_3, C_2H_5) Cl^- → C_2H_5O—C=O (with CH_3) HCl

There is 1 mark for each of the curly arrows in the first stage, but you must show one lone pair on the oxygen atom of the alcohol to gain the second mark. For the third mark you must show the correct structure of the intermediate, the correct charges and two correct curly arrows. The fourth mark is for showing the loss of the H^+, with the curly arrow showing the electron pair returning to the oxygen.

(iii) It is not an equilibrium, so gives a better yield ✓.

Esterifications involving carboxylic acids and alcohols are equilibrium reactions, requiring heat and a catalyst. Acyl chlorides are more reactive than carboxylic acids because they possess two electron-withdrawing groups (C=O and C–Cl), making the carbon electron deficient and susceptible to nucleophilic attack.

(b) (i) $CH_3CH_2COCl + 2C_2H_5NH_2 \longrightarrow CH_3CH_2CONHC_2H_5 + C_2H_5NH_3^+Cl^-$ ✓ ✓

This equation is more difficult than that involving the alcohol because the amine reacts with the liberated hydrogen chloride to produce a salt. The following equation would score 1 mark:

$CH_3CH_2COCl + C_2H_5NH_2 \longrightarrow CH_3CH_2CONHC_2H_5 + HCl$

(ii)

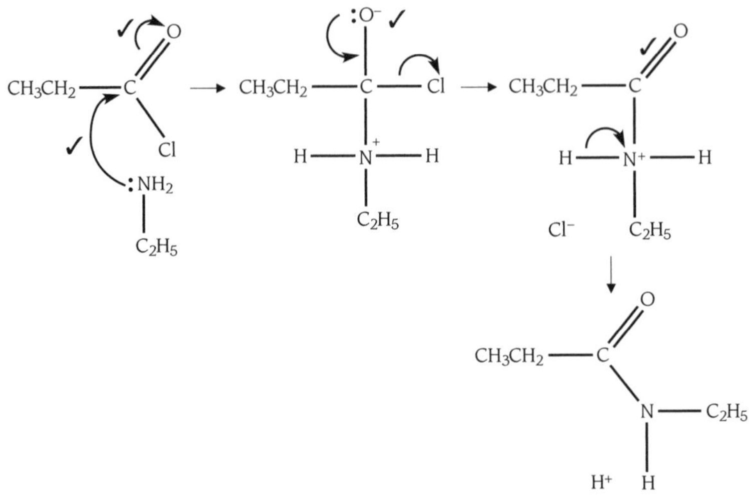

CH_3CH_2—C(=O)—Cl ✓ ✓ with :NH_2—C_2H_5 $\longrightarrow$ CH_3CH_2—C(:O^- ✓)(—Cl)—$\overset{+}{N}$(H)(H)—C_2H_5 $\longrightarrow$ CH_3CH_2—C(=O ✓)—N^+(H)(H)—C_2H_5 Cl^-

$\downarrow$

CH_3CH_2—C(=O)—N(H)—C_2H_5 H^+

The mechanism is similar to the reaction of an acyl chloride with an alcohol. The only difference is that the $-NH_2$ group replaces the $-OH$ group.

(iii) X: $CH_3CH_2CH_2$—C(=O)—Cl ✓ H—N(H)—C_6H_5 ✓

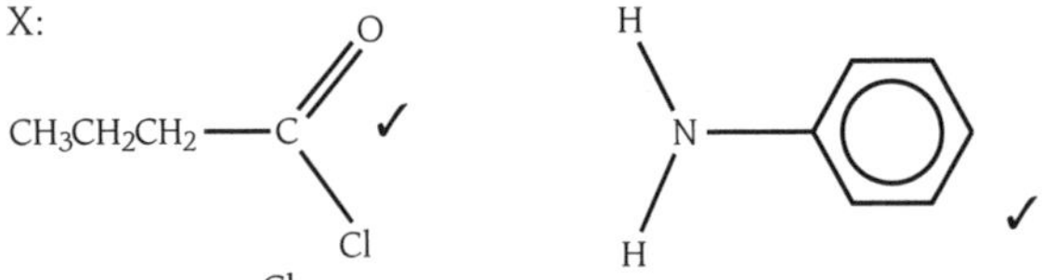

Y: C_6H_5—C(=O)—Cl ✓ H—N(H)—CH(CH_3)(CH_3) ✓

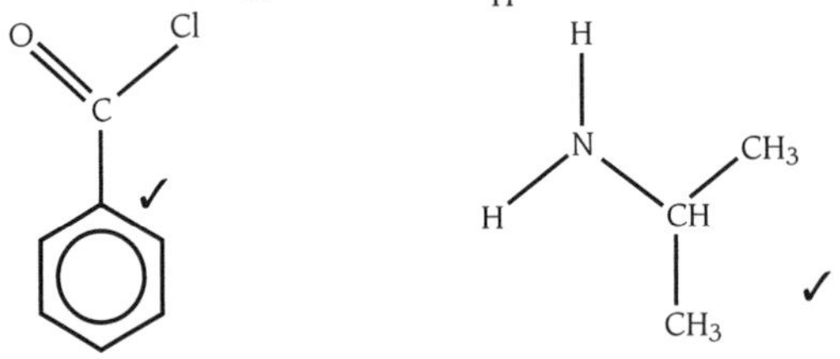

To predict the acyl chlorides and amines used, split the amide at the C–N bond and add Cl to the carbon of the C=O group and H to the N of the NH group.

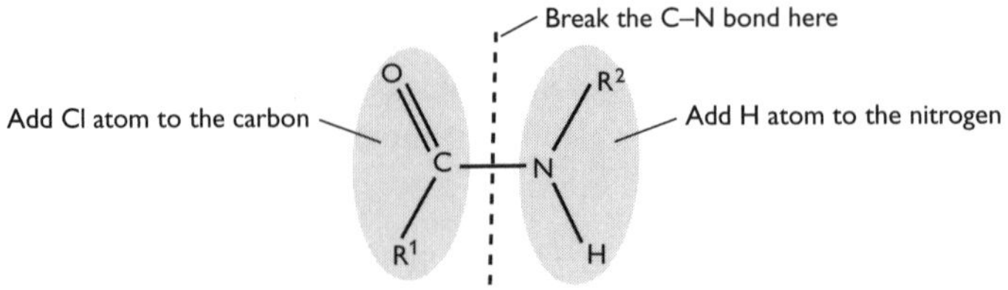

(c) (i)

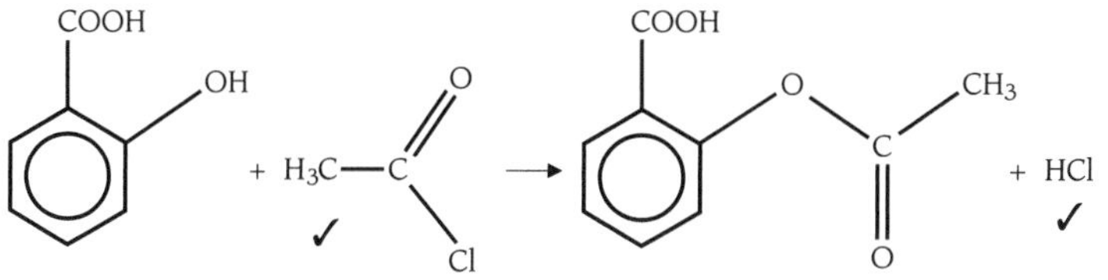

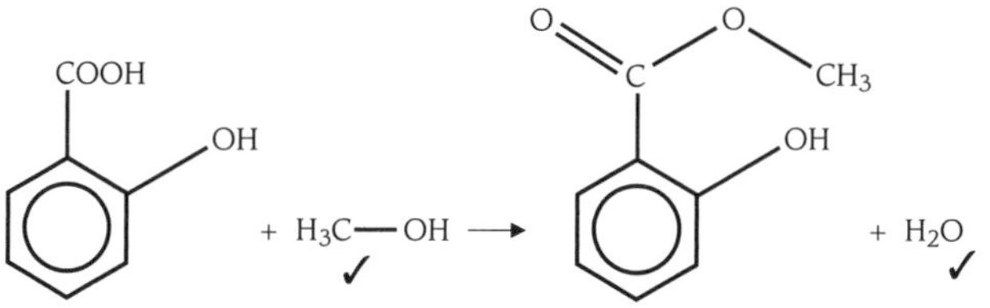

COOH ... OH + H_3C—C(=O)Cl ✓ → COOH ... O—C(=O)—CH_3 + HCl ✓

COOH ... OH + H_3C—OH ✓ → O=C(—O—CH_3) ... OH + H_2O ✓

2-hydroxybenzenecarboxylic acid possesses both an –OH group and a –COOH group. The –OH group reacts with an acyl chloride (CH_3COCl) to produce an ester. The –COOH group reacts with the alcohol (CH_3OH) to produce a different ester. A common mistake is to omit the HCl or H_2O from the equation.

(ii) Concentrated sulfuric acid ✓

Carboxylic acids are less reactive than acyl chlorides. Esterification requires a catalyst, unlike acylation.

(iii) Ethanoic anhydride ✓

An alternative reagent to CH_3COCl is ethanoic anhydride and the equation is:

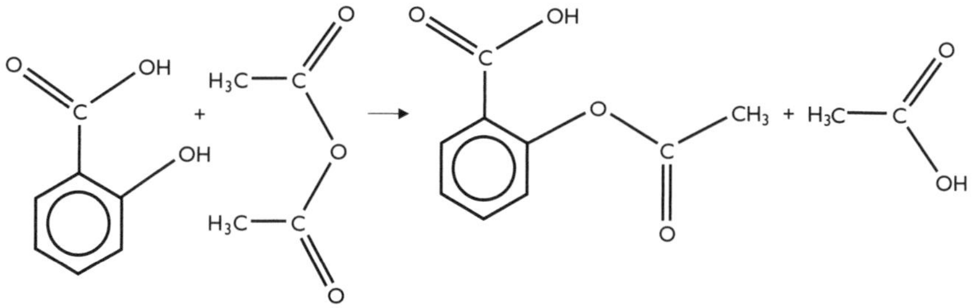

O=C(OH) ... OH + H_3C—C(=O)—O—C(=O)—CH_3 → O=C(OH) ... O—C(=O)—CH_3 + H_3C—C(=O)OH

Ethanoic anhydride is the common reagent used in industry. If you have used this compound in part (i) then the alternative reagent in this section must be CH_3COCl.

(d) C ✓, B ✓, B ✓

All these compounds will react with NaOH but only compound C will undergo hydrolysis to produce $C_6H_5COO^-Na^+$ and CH_3OH. Compound B will undergo nucleophilic addition–elimination reactions with ammonia to produce $C_6H_5CONH_2$ and NH_4Cl. Compound B is an acyl chloride, so it reacts with cold water in a nucleophilic addition–elimination reaction to produce C_6H_5COOH and HCl.

Aromatic chemistry

(a) The enthalpy of hydrogenation of compound X is $\Delta H = -120\,kJ\,mol^{-1}$.

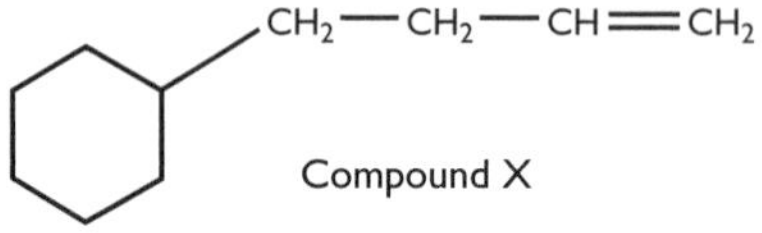

$CH_2-CH_2-CH=CH_2$

Compound X

(i) Use this value to predict the enthalpy of hydrogenation of compounds A to D below. In compounds C and D it is assumed that there are alternate carbon–carbon single bonds and carbon–carbon double bonds in the cyclic part of the molecule. (4 marks)

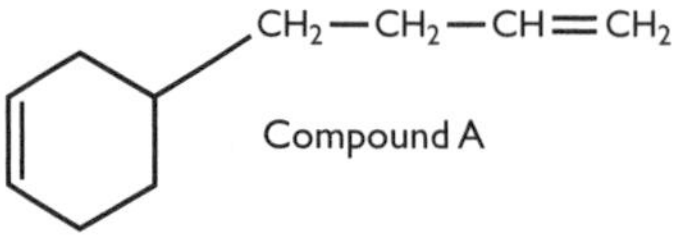

$CH_2-CH_2-CH=CH_2$

Compound A

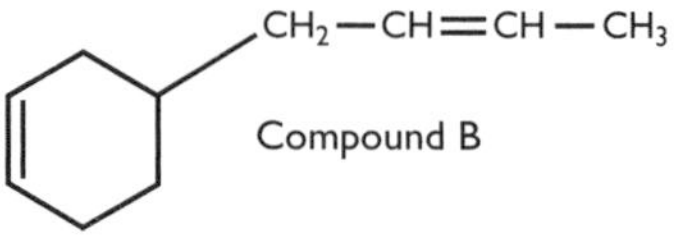

$CH_2-CH=CH-CH_3$

Compound B

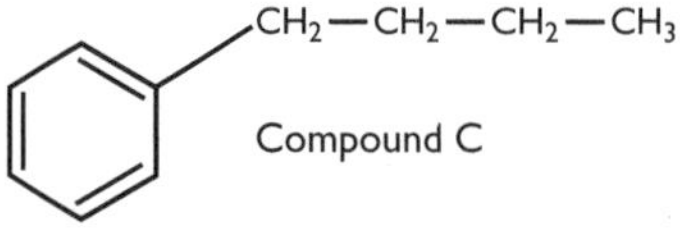

$CH_2-CH_2-CH_2-CH_3$

Compound C

$CH_2-CH_2-CH=CH_2$

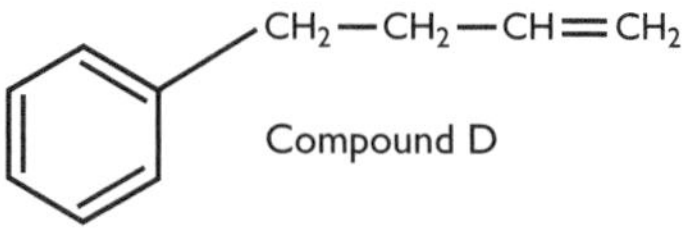

Compound D

(ii) The structure of 4-phenylbut-1-ene is shown below:

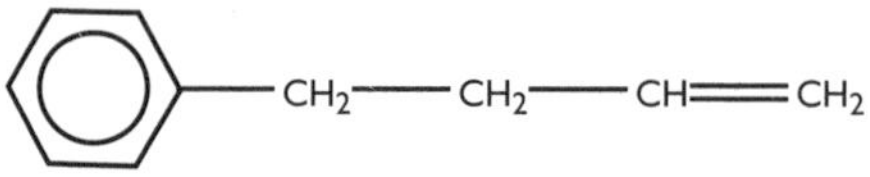

$CH_2-CH_2-CH=CH_2$

Explain why the enthalpy of hydrogenation of 4-phenylbut-1-ene is approximately $-328\,kJ\,mol^{-1}$. (2 marks)

(b) The structure of compound Y is shown below. It is formed in the nitration of 4-phenylbut-1-ene.

O_2N- $CH_2-CH_2-CH=CH_2$

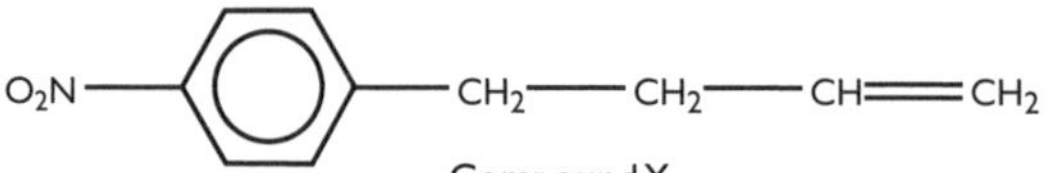

Compound Y

(i) State the reagents needed for nitration. (2 marks)

(ii) Write an equation for the formation of compound Y from 4-phenylbut-1-ene. (1 mark)

(iii) Write a series of equations to show the formation of the nitronium ion (nitryl cation) during the nitration process. (3 marks)

(iv) Outline the mechanism for the nitration of 4-phenylbut-1-ene. (4 marks)

(v) Draw the structure of another possible product in the mononitration of 4-phenylbut-1-ene. (1 mark)

(c) (i) State the reagents required to convert benzene into the compound shown below.

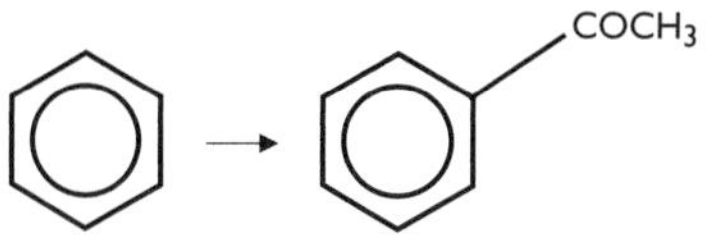

COCH$_3$

(2 marks)

(ii) Write an equation for the formation of the electrophile. (1 mark)

(iii) Outline the mechanism for the reaction. (3 marks)

Total: 23 marks

Grade-A answer to Question 8

(a) (i) Compound A = –240 kJ mol^{-1} ✓; Compound B = –240 kJ mol^{-1} ✓; Compound C = –360 kJ mol^{-1} ✓; Compound D = –480 kJ mol^{-1} ✓

One C=C bond liberates –120 kJ mol^{-1} upon hydrogenation. Two C=C bonds would be expected to liberate –240 kJ mol^{-1}, three C=C bonds would liberate –360 kJ mol^{-1} and four C=C bonds would liberate –480 kJ mol^{-1}. This assumes there is no delocalisation of electrons in the structures.

(ii) The 4-phenylbut-1-ene should have an enthalpy of hydrogenation of –480 kJ mol^{-1} but it is 152 kJ mol^{-1} more stable than expected ✓. This is because the alternate single and double bonds in the benzene ring exist as a ring of delocalised electrons, which causes this extra stability ✓.

In 4-phenylbut-1-ene there are four double bonds, three of which are in the benzene. However, the electrons in the benzene form a ring, or cloud, of delocalised electrons. This makes the compound more stable than expected, so the enthalpy of hydrogenation liberates less energy than anticipated, that is, 480 – 328 = 152 kJ mol^{-1}. This difference is called the delocalisation energy.

(b) (i) Concentrated nitric acid ✓ and concentrated sulfuric acid ✓.

You must include the term 'concentrated' for both reagents.

(ii)

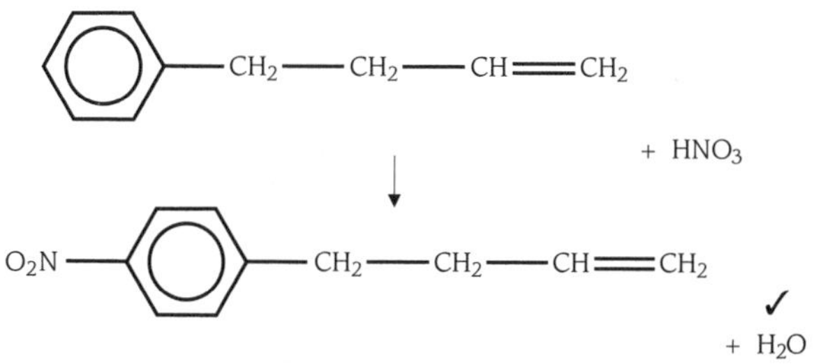

C_6H_5—CH$_2$—CH$_2$—CH=CH$_2$ + HNO$_3$

↓

O$_2$N—C_6H_4—CH$_2$—CH$_2$—CH=CH$_2$ + H$_2$O ✓

H_2SO_4 is not included in the final equation. A common mistake is to omit the H_2O.

(iii) $HNO_3 + H_2SO_4 \longrightarrow H_2NO_3^+ + HSO_4^-$ ✓
$H_2NO_3^+ \longrightarrow H_2O + NO_2^+$ ✓
$H_2SO_4 + H_2O \longrightarrow HSO_4^- + H_3O^+$ ✓

The H_2SO_4 acts as a strong acid and protonates the HNO_3, which decomposes readily to produce the NO_2^+ electrophile.

(iv)

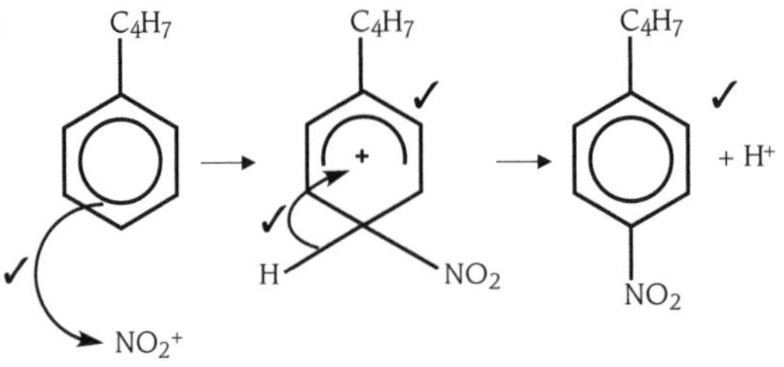

C_4H_7 C_4H_7 C_4H_7
✓ ✓ ✓ + H^+
✓ H NO_2 NO_2
NO_2^+

There are 4 marks for the mechanism. The first mark is for the curly arrow from the ring of delocalised electrons to the NO_2^+ ion, the second mark is for the structure of the intermediate, which must show the $-NO_2$ at the 4 position and the broken ring of electrons in the correct position. The third mark is for the curly arrow showing the hydrogen leaving and the electron pair from the bond returning to re-form the complete ring of delocalised electrons. The final mark is for the product, with a complete ring of delocalised electrons, and the H^+.

(v)

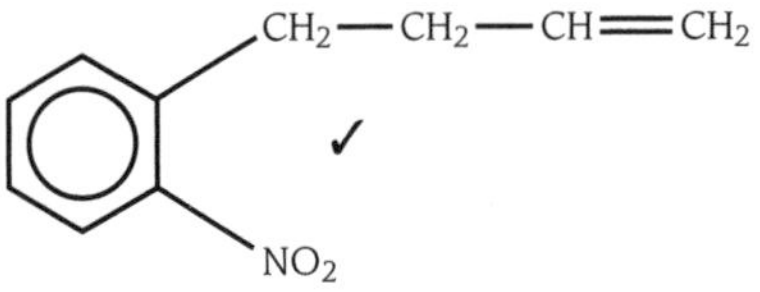

$CH_2—CH_2—CH{=}CH_2$
✓
NO_2

An alternative is to show the $-NO_2$ group at the 3 position.

(c) (i) CH_3COCl ✓ and $AlCl_3$ ✓
(ii) $CH_3COCl + AlCl_3 \longrightarrow {}^+COCH_3 + AlCl_4^-$ ✓

The $AlCl_3$ acts as a Lewis acid catalyst and accepts a lone pair to form $AlCl_4^-$.

(iii)

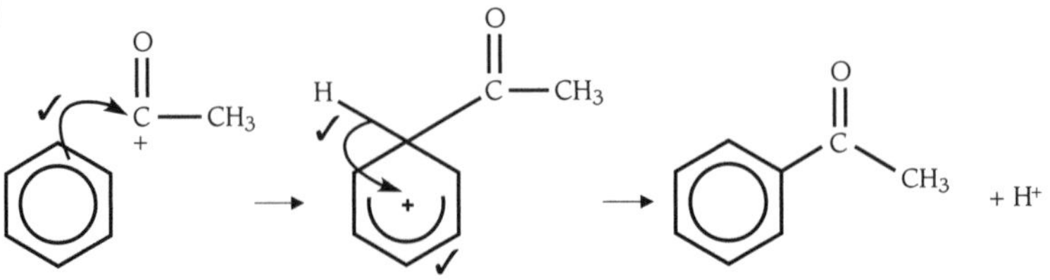

O O O
✓ C — CH_3 H C — CH_3 C
\+ ✓ CH_3 + H^+
✓

The mechanism follows the same principle as the nitration in part **(b) (iv)** with 1 mark for each curly arrow and 1 mark for the intermediate structure. It is essential that you are accurate when drawing the broken ring of delocalised electrons.

Amines

(a) Write an equation for the reaction of propylamine with (i) water and (ii) hydrochloric acid. Explain why propylamine is acting as a base in these reactions. (3 marks)

(b) Explain why propylamine is a stronger base than ammonia. (2 marks)

(c) Consider the reaction schemes below, which summarise the two methods available for the preparation of propylamine.

Method A

$$H_3C—CH_2—Br \xrightarrow{\text{Step 1}} H_3C—CH_2—CN \xrightarrow{\text{Step 2}} H_3C—CH_2—CH_2—NH_2$$

Method B

$$H_3C—CH_2—CH_2—Br \xrightarrow{\text{One step}} H_3C—CH_2—CH_2—NH_2$$

(i) Classify the reaction type in:
- **method A, step 1**
- **method A, step 2**
- **method B** (3 marks)

(ii) Write an equation for:
- **method A, step 1**
- **method A, step 2**
- **method B** (3 marks)

(iii) What is the main advantage of step 2 in method A over the single-step method B in the preparation of propylamine? (2 marks)

(iv) Draw the mechanism for the formation of propylamine in method B. (4 marks)

(v) Give the structures of three other possible nitrogen-containing products when propylamine is prepared from 1-bromopropane in method B. (3 marks)

(d) Consider the reaction sequence outlined below.

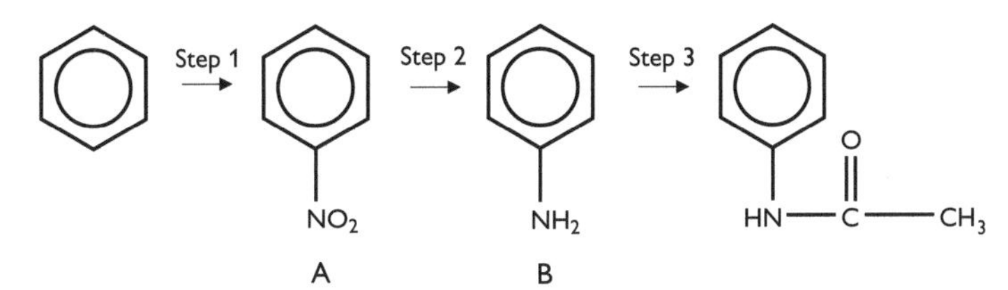

(i) Explain why compound B is less basic than ammonia. (2 marks)

(ii) Classify the reaction type for each of the steps 1, 2 and 3. (3 marks)

(iii) Write an equation to describe each of the steps 1, 2 and 3. (3 marks)

(iv) State the reagents needed in step 2. (1 mark)

(v) State the role of compound B in step 3. (1 mark)

Total: 30 marks

Grade-A answer to Question 9

(a) (i) $CH_3CH_2CH_2NH_2 + H_2O \rightleftharpoons CH_3CH_2CH_2NH_3^+ + OH^-$ ✓

(ii) $CH_3CH_2CH_2NH_2 + HCl \longrightarrow CH_3CH_2CH_2NH_3^+ + Cl^-$ ✓

The propylamine accepts a proton (or H^+ ion) ✓ and so is acting as a Brønsted–Lowry base.

(b) The alkyl group $(CH_3CH_2CH_2)$– is electron releasing ✓. This makes the lone pair on the nitrogen more readily available ✓.

Various expressions are used to describe the electron-releasing effect of the alkyl group. The alkyl group 'has a positive inductive effect' or 'pushes electrons towards the nitrogen' are acceptable answers for the first mark.

(c) (i) Method A, step 1 is nucleophilic substitution ✓.
Method A, step 2 is reduction ✓.
Method B is nucleophilic substitution ✓.

Step 2 in method A could be classified as hydrogenation.

(ii) $CH_3CH_2Br + KCN \longrightarrow CH_3CH_2CN + KBr$ ✓
$CH_3CH_2CN + 2H_2 \longrightarrow CH_3CH_2CH_2NH_2$ ✓
$CH_3CH_2CH_2Br + 2NH_3 \longrightarrow CH_3CH_2CH_2NH_2 + NH_4Br$ ✓

In the first equation, NaCN could be used but not HCN. In the second, the symbol 4[H] could be used to represent the reducing agent. In the third, $2NH_3$ is essential because the liberated HBr reacts with the NH_3 to produce the salt NH_4Br.

(iii) In step 2 of method A there is only one product ✓ whereas in method B there is a risk of further substitution to produce a mixture of amines and the quaternary ammonium salt ✓.

The amines produced as a result of nucleophilic substitution still possess a lone pair. They can act as nucleophiles and attack the haloalkane, resulting in a mixture of products. The reaction stops when the quaternary ammonium salt is produced because the nitrogen has four groups attached and there is no lone pair available.

(iv)

$H_3C—CH_2—CH_2—Br$ ✓ ✓ $:NH_3$ → $H_3C—CH_2—CH_2—N^+H_3$ Br^- ✓ ✓ $:NH_3$
↓
$H_3C—CH_2—CH_2—NH_2$
NH_4^+

There are 3 marks for the curly arrows and 1 mark for the intermediate structure. A common mistake is to omit the lone pair from the nitrogen in ammonia.

(v) $(CH_3CH_2CH_2)_2NH$ ✓, $(CH_3CH_2CH_2)_3N$ ✓, $(CH_3CH_2CH_2)_4N^+Br^-$ ✓

The amines continue to attack the original haloalkane to produce a mixture of secondary and tertiary amines. The final product is a quaternary ammonium salt.

(d) (i) The lone pair on the nitrogen atom in the $-NH_2$ is less available ✓ in compound B because it interacts with the delocalised electrons in the benzene ring ✓.

Basicity is determined by the availability of the lone pair. Always approach it from this angle, rather than the ability to accept a proton.

(ii) Step 1 = nitration ✓, step 2 = reduction ✓, step 3 = acylation ✓

The question asks for reaction *types* not names of mechanisms.

(iii)

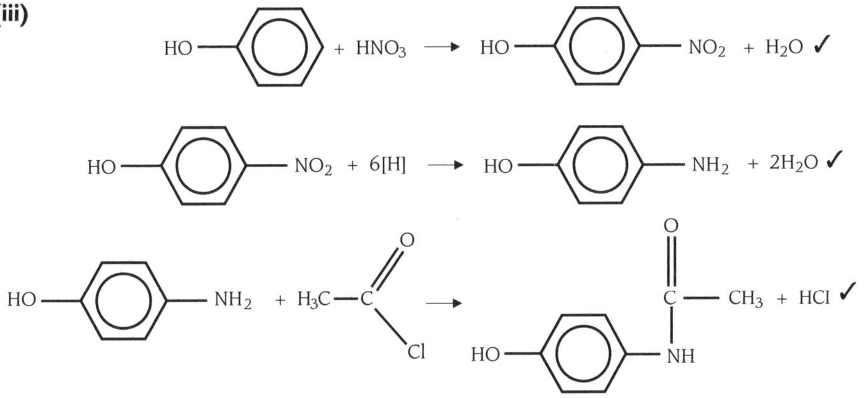

$$HO-C_6H_4 + HNO_3 \rightarrow HO-C_6H_4-NO_2 + H_2O \checkmark$$

$$HO-C_6H_4-NO_2 + 6[H] \rightarrow HO-C_6H_4-NH_2 + 2H_2O \checkmark$$

$$HO-C_6H_4-NH_2 + H_3C-C(=O)Cl \rightarrow HO-C_6H_4-NH-C(=O)-CH_3 + HCl \checkmark$$

Concentrated sulfuric acid is used in nitration to generate NO_2^+ but is not included in the overall equation. A common mistake is to omit the H_2O. The alternative to 6[H] is $3H_2$ in the reduction of the NO_2 group. The alternative to ethanoyl chloride is ethanoic anhydride, which forms CH_3COOH rather than HCl.

(iv) Tin and concentrated hydrochloric acid ✓.

Alternative reagents in step 2 are hydrogen and a nickel catalyst.

(v) Compound B is acting as a nucleophile ✓.

Compound B has a lone pair on the nitrogen of the $-NH_2$ group and donates it to the electron-deficient carbon of the ethanoyl chloride, so it is acting as a nucleophile.

Question 10

Amino acids

(a) The structure of phenylalanine is shown below:

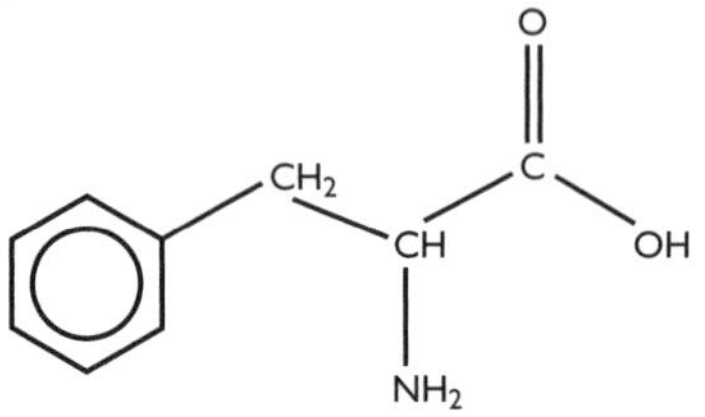

(i) Draw the structure of phenylalanine when it is in a solution of
- **sodium hydroxide**
- **hydrochloric acid** (2 marks)

(ii) Draw the structure of phenylalanine at its isoelectric point. (1 mark)

(iii) Phenylalanine is isomeric with the compound benzocaine.

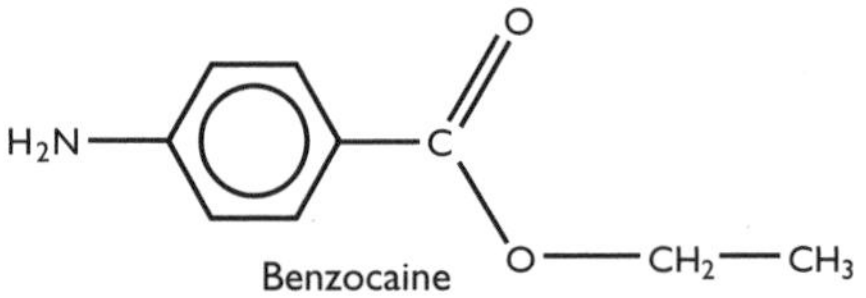

Benzocaine

State which compound has the higher melting point. Explain why. (4 marks)

(iv) Draw the structure of the amino acid 2-aminopropanoic acid. Draw the structure of the dipeptide formed when phenylalanine reacts with 2-aminopropanoic acid and circle the peptide link in the dipeptide. (3 marks)

(v) Predict the structure of the organic product when phenylalanine reacts with ethanoyl chloride. Give the formula of the other product of the reaction. (2 marks)

(b) Consider the tripeptide shown below:

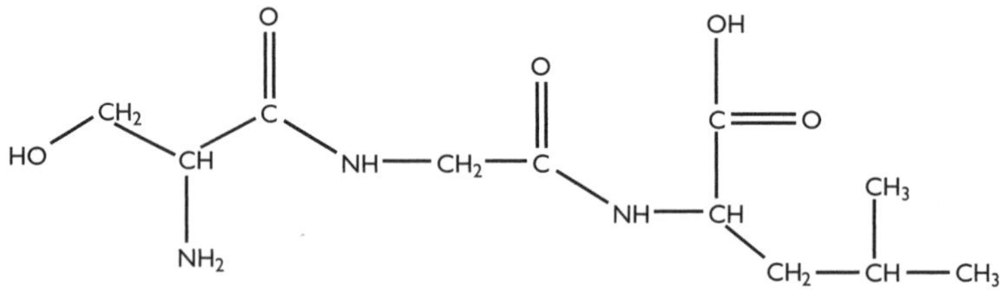

(i) Copy the structure of the tripeptide then label each chiral centre present in the tripeptide with an asterisk (*). Circle the peptide links. (3 marks)

(ii) State the type of intermolecular force that holds together molecules of the tripeptide. (1 mark)

(iii) Name a reagent that would hydrolyse the tripeptide. Draw the structure of the constituent amino acids that would result from the hydrolysis. (4 marks)

(iv) What laboratory technique is suitable to separate the constituent amino acids formed in part (iii)? Outline how the amino acids could be identified using this technique. (4 marks)

Total: 24 marks

■ ■ ■

Grade-A answer to Question 10

(a) (i) The structure in NaOH will be:

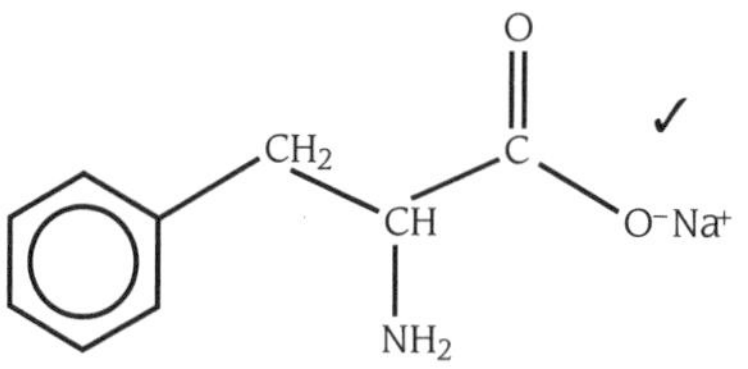

The structure in HCl will be:

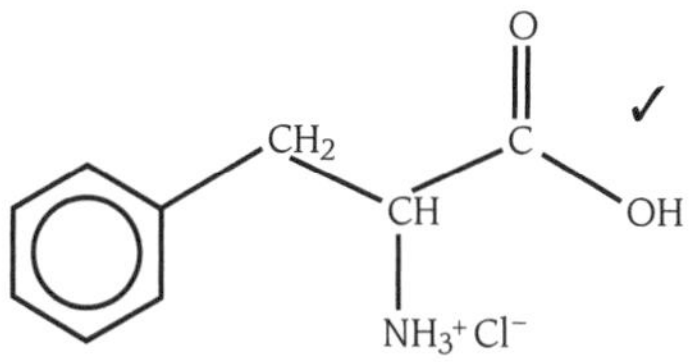

Amino acids are both basic (due to the $-NH_2$ group) and acidic (due to the $-COOH$ group). When predicting the reactions of amino acids, treat the two functional groups separately. Addition of an alkali leads to a reaction with the acid part of the molecule to produce a salt of the type $H_2NRCOO^-Na^+$. Addition of an acid leads to a reaction with the basic part of the molecule to produce a salt of the type $HOOCRNH_3^+Cl^-$.

(ii) The structure at the isoelectric point will be the zwitterion.

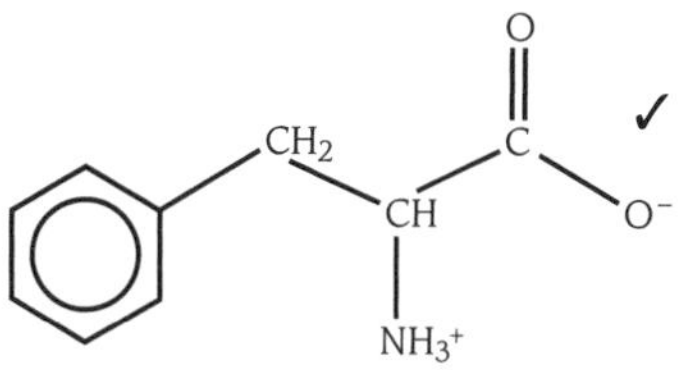

At the isoelectric point, a zwitterion exists because the acid part of the molecule has donated an H^+ ion to the basic part of the molecule, to give COO^- and NH_3^+.

(iii) Phenylalanine will have the higher melting point ✓ because it will exist as zwitterions ✓, with strong electrostatic forces of attraction between oppositely charged ions ✓. Benzocaine will have hydrogen bonding ✓.

The first mark is for the correct initial statement. If this is incorrect, no further marks can be awarded.

(iv)

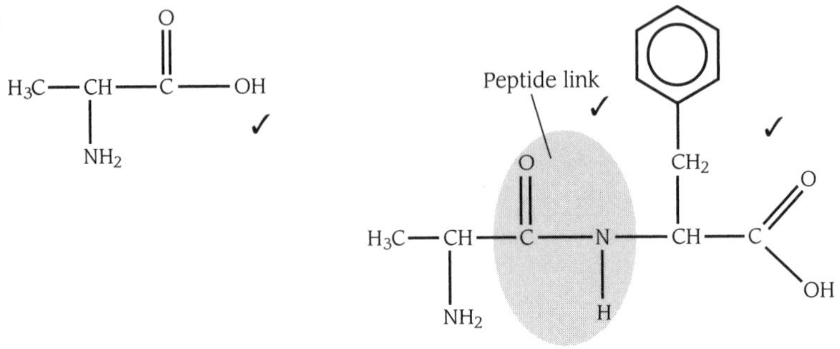

$H_3C—CH(NH_2)—C(=O)—OH$ ✓

Peptide link ✓

$H_3C—CH(NH_2)—C(=O)—N(H)—CH(CH_2C_6H_5)—C(=O)OH$ ✓

An alternative structure is:

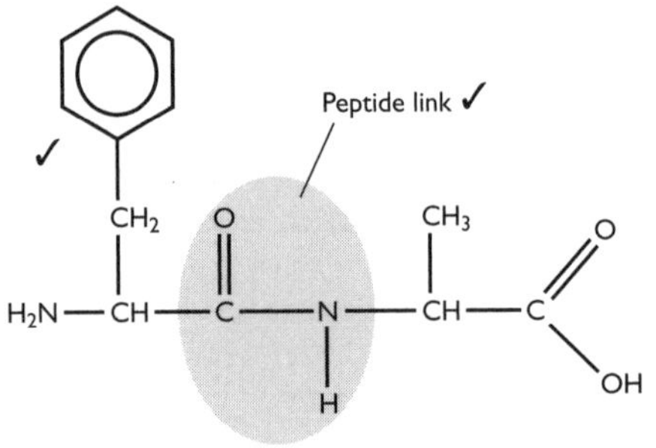

Peptide link ✓

$H_2N—CH(CH_2C_6H_5)—C(=O)—N(H)—CH(CH_3)—C(=O)OH$ ✓

(v)

$C_6H_5CH_2—CH(NH—C(=O)CH_3)—C(=O)OH$ ✓ HCl ✓

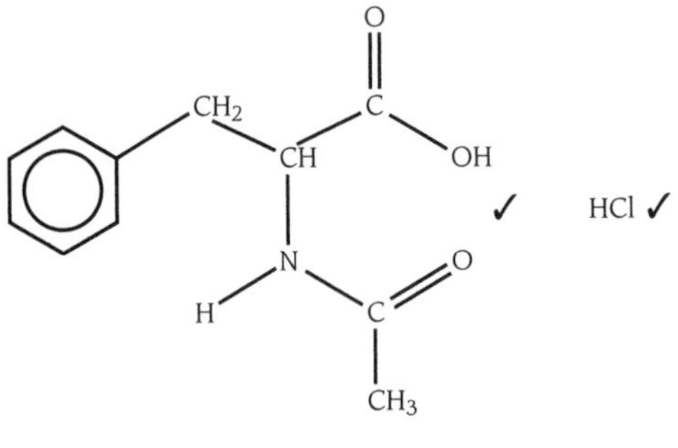

In this reaction, you are using your knowledge of amine chemistry to predict the product. The lone pair on the nitrogen of the $–NH_2$ group will attack the electron-deficient carbon atom of the ethanoyl chloride.

(b) (i)

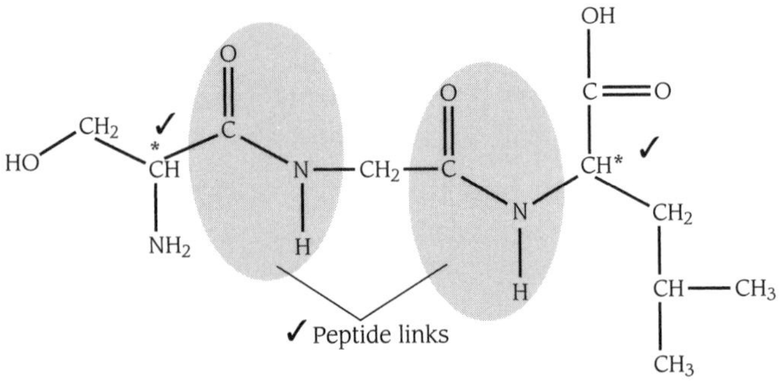

$HO—CH_2—\overset{*}{C}H(NH_2)—C(=O)—N(H)—CH_2—C(=O)—N(H)—CH^*(COOH)—CH_2—CH(CH_3)—CH_3$ ✓ ✓

✓ Peptide links

(ii) Hydrogen bonding ✓

(iii) Sodium hydroxide ✓; the constituent amino acids would be:

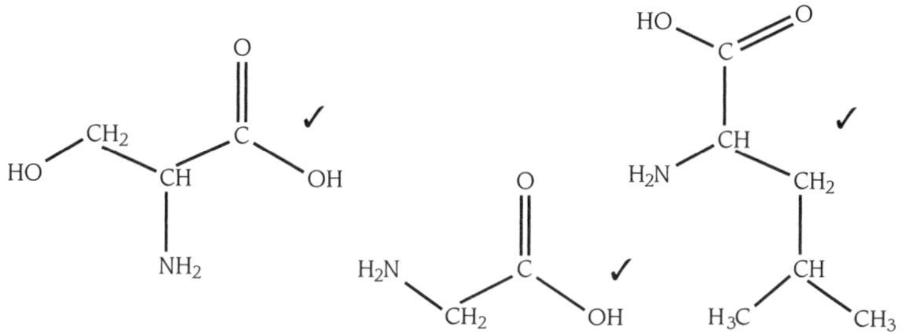

HO–CH₂–CH(NH₂)–C(=O)–OH ✓   H₂N–CH₂–C(=O)–OH ✓   HO–C(=O)–CH(H₂N)–CH₂–CH(H₃C)–CH₃ ✓

Looking for carbon atoms with four different groups attached identifies the chiral centres. There are 2 marks in part (i) for the two chiral centres. A peptide link is a –CONH– group and identifying both peptide links will gain the third mark. The intermolecular forces are hydrogen bonds, owing to the presence of the –NH– and C=O groups. The peptide links are susceptible to acid (e.g. with HCl) or alkaline hydrolysis and will split at the C–N bond to regenerate the original amino acids. If alkaline hydrolysis is used, then the amino acids form salts of the type $H_2NRCOO^-Na^+$. If acid hydrolysis is used, then the amino acids form salts of the type $Cl^-H_3N^+RCOOH$, so alternatives to this grade-A answer are acceptable.

(iv) Paper chromatography *or* thin-layer chromatography ✓. Spray with ninhydrin ✓ measure the R_f values for each of the spots ✓. Compare with the R_f values of amino acids in a data book ✓.

Another way to gain the last two marks would be to run samples of known amino acids on the same chromatogram as the mixture and compare their positions with those of the components of the mixture.

Polymers

(a) (i) **Draw the repeating units and name the addition polymers formed from the following monomers:**
- **propene**
- **chloroethene** (4 marks)

(ii) **Explain why neither of these polymers is biodegradable.** (2 marks)

(iii) **Give one advantage and one disadvantage of disposing of articles made from the polymer of chloroethene by incineration.** (2 marks)

(b) **Nylon-6,10 is formed from the monomers decanedioyl chloride and hexane-1,6-diamine.**

(i) **Draw the structures of decanedioyl chloride and hexane-1,6-diamine.**

(ii) **Draw the structure of the repeating unit of nylon-6,10 and identify the small molecules lost in the formation of the polymer.** (4 marks)

(c) **Draw the repeating units formed by the following amino acids:**

(i) Glycine

$H_2N-CH_2-C(=O)-OH$

(ii) Leucine

$H_3C-CH(CH_3)-CH_2-CH(NH_2)-C(=O)-OH$

(2 marks)

(d) **Consider the following polymer chains labelled A to C:**

A $-C(=O)-C_6H_4-C(=O)-OCH_2CH_2O-C(=O)-C_6H_4-C(=O)-OCH_2CH_2O-$

B $-CH_2-CH(C_6H_5)-CH_2-CH(C_6H_5)-CH_2-CH(C_6H_5)-CH_2-CH(C_6H_5)-$

C $-NH(CH_2)_6NH-C(=O)-(CH_2)_4-C(=O)-NH(CH_2)_6NH-C(=O)-(CH_2)_4-C(=O)-$

Using the letters A to C, identify the polymer or polymers:

(i) **formed by addition polymerisation**

(ii) **formed by condensation polymerisation**

(iii) **held together by van der Waals forces only**

(iv) **held together by hydrogen bonding**

(v) **which are biodegradable** (7 marks)

(e) Kevlar is an aromatic polyamide formed from the reaction of benzene-1,4-dicarboxylic acid and benzene-1,4-diamine. Kevlar has very high tensile strength and is used for lightweight bullet-proof vests.

(i) Draw the structures of the two monomers used in this reaction. (2 marks)

(ii) Draw the structure of the repeating unit of Kevlar. (1 mark)

(iii) Explain why Kevlar has great strength. (3 marks)

Total: 27 marks

■ ■ ■

Grade-A answer to Question 11

(a) (i)

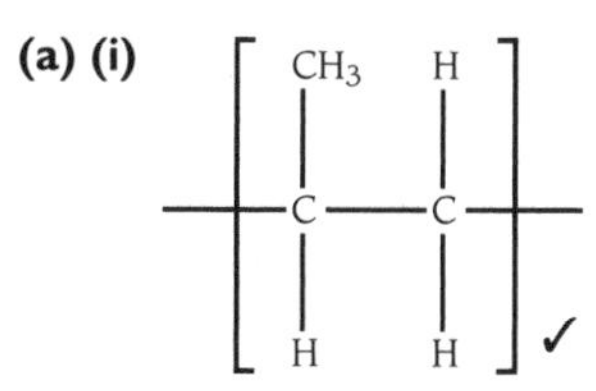

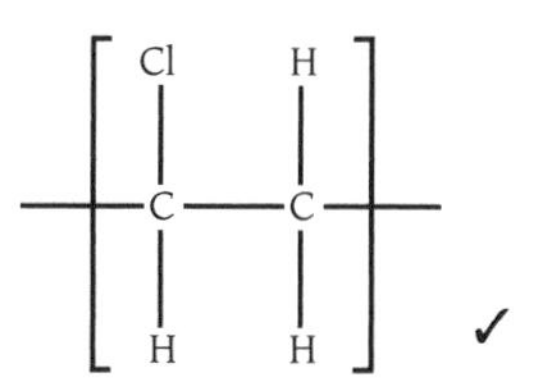

Poly(propene) ✓ Poly(chloroethene) ✓

Draw out the structures fully so that the positions of the methyl group in poly(propene) and of the chlorine atom in poly(chloroethene) are clear. A common mistake is to show poly(propene) as a three-carbon repeating unit with no branching, ~~$-(CH_3CHCH_2)-$~~.

(ii) Both are chemically inert ✓ and also cannot be broken down by enzymes ✓.

(iii) Advantage: reduces the amount sent to landfill *or* source of energy ✓; disadvantage: gives off HCl fumes ✓.

(b) (i) $ClOC-CH_2-CH_2-CH_2-CH_2-CH_2-CH_2-CH_2-CH_2-COCl$ ✓

$H_2N-CH_2-CH_2-CH_2-CH_2-CH_2-CH_2-NH_2$ ✓

(ii) $-[-CO-(CH_2)_8-CO-NH-(CH_2)_6-NH-]-$ ✓ HCl ✓

The most common mistake in this type of question is to include too many carbon atoms in the decanedioyl chloride. Remember that the –COCl groups at each end of the molecule provide two of the carbons, so only eight CH_2 groups are required. When drawing the repeating unit of the condensation polymer, it is important that the covalent bonds go through the brackets. If you have drawn

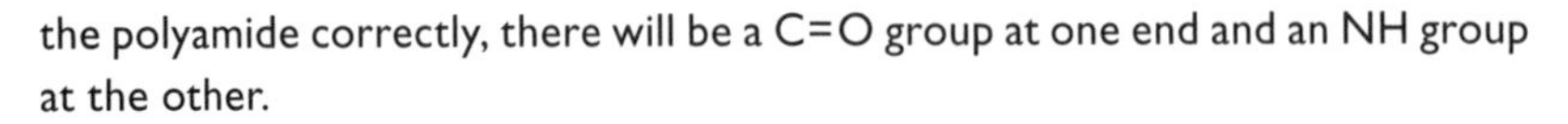

the polyamide correctly, there will be a C=O group at one end and an NH group at the other.

(c) **(i)**

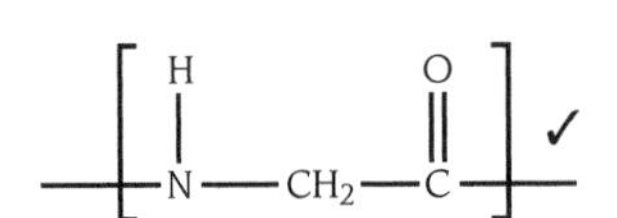

(ii) ✓

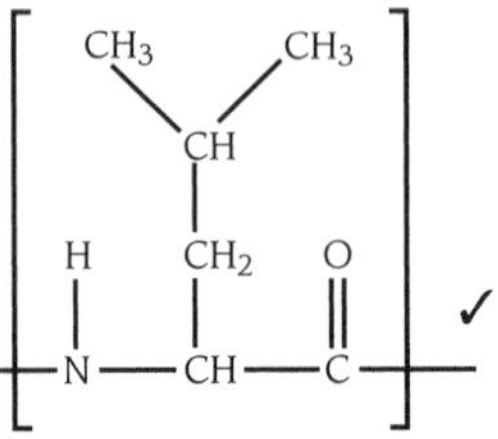

Amino acids undergo self-polymerisation. Always draw the molecule with the $-NH_2$ group at one end and the –COOH at the other. To convert it into the polymer, take H from the $-NH_2$ and OH from the –COOH.

(d) **(i)** B ✓

(ii) A ✓ and C ✓

(iii) B ✓

(iv) C ✓

(v) A ✓ and C ✓

Polyalkenes are non-polar hydrocarbon chains which attract each other by van der Waals forces. Polyesters are polar due to the COO group and attract each other by dipole–dipole forces. Polyamides contain NH and C=O groups so they attract each other by hydrogen bonding. Polyesters and polyamides are susceptible to attack at their linkages so they can be easily broken down, whereas polyalkenes are non-polar, inert and non-biodegradable.

(e) **(i)** ✓ ✓

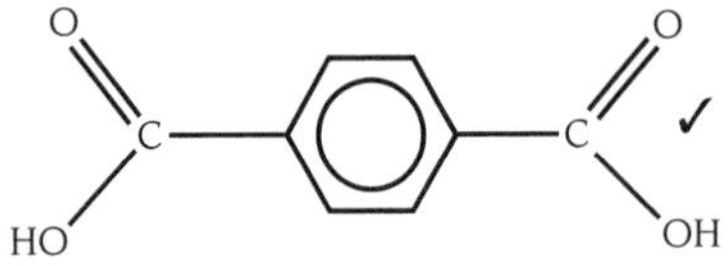

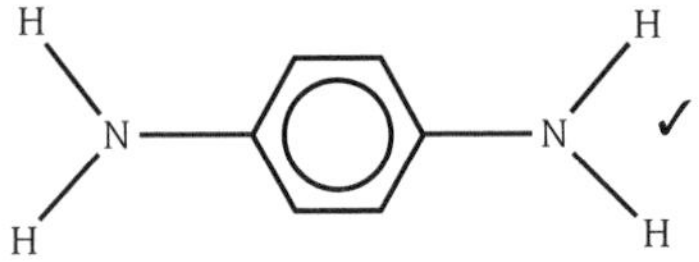

(ii) ✓

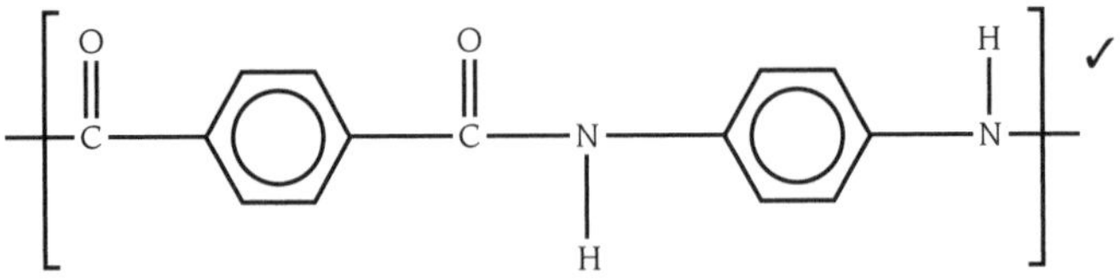

(iii) Hydrogen bonding ✓ between the N–H and C=O groups ✓ along adjacent linear chains ✓.

The structures of the monomers are relatively easy to predict from their names. In this reaction, the OH is lost from the –COOH groups and the H is lost from the $-NH_2$ group to produce the polymer and molecules of water.

Organic synthesis and analysis

(a) The pain-killing drug paracetamol can be prepared using the following synthetic route:

Phenol (OH on benzene) —Step 1→ (OH, NO_2) —Step 2→ (OH, NH_2) —Step 3→ Paracetamol (OH, HN—C(=O)—CH_3)

(i) State the reaction types in steps 1, 2 and 3. (3 marks)

(ii) State the reagents used in steps 1, 2 and 3. (3 marks)

(iii) State the name of and outline the mechanism for the reaction in step 1. (4 marks)

(b) Lactic acid can be formed from ethanol using the following synthetic route:

$H_3C—CH_2—OH$ —Step 1→ $H_3C—C(=O)H$ (**A**) —Step 2→ $H_3C—C(OH)(H)—C{\equiv}N$ (**B**) —Step 3→ $H_3C—C(OH)(H)—C(=O)OH$ (Lactic acid)

(i) Name the intermediates A and B. (2 marks)

(ii) State the reaction types in steps 1 and 3. (2 marks)

(iii) State the name of, and outline the mechanism for, step 2. (4 marks)

(iv) State the reagents and essential conditions for step 1. (3 marks)

(v) State the reagents for steps 2 and 3. (2 marks)

(c) Describe how you could distinguish between the following pairs of organic compounds using a simple chemical test. Name the reagents and describe the expected observations. If there is no reaction, then state 'no reaction'.
If a reaction takes place, include a balanced equation.

(i)

1: $C_6H_5—C(=O)—CH_2—CH_3$ and

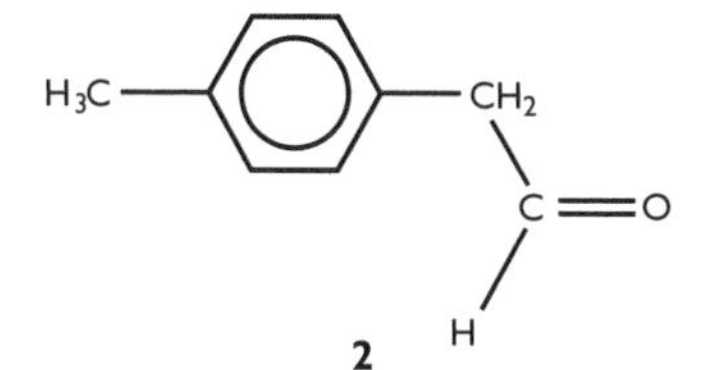

(ii)

3: $H_3C—C_6H_{10}—CH(CH_3)_2$ and **4**: $H_3C—C_6H_8—C(CH_3)=CH_2$

(iii) 5 and 6 (12 marks)

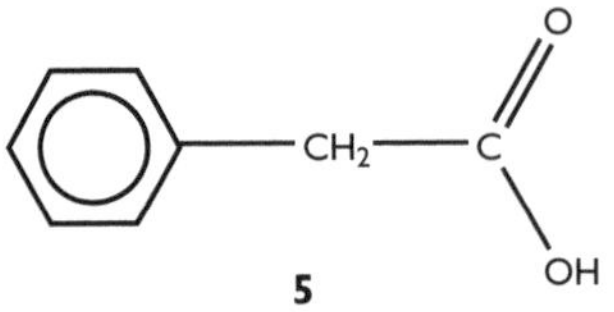

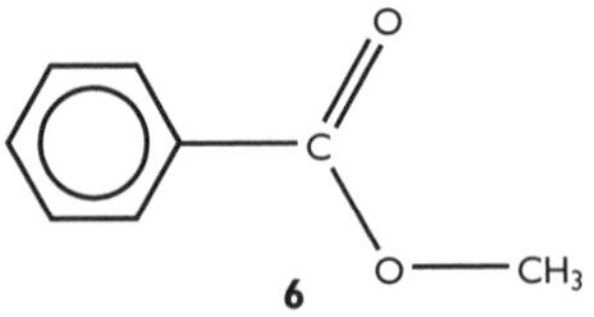

(d) The synthesis of the anaesthetic drug benzocaine is shown below.

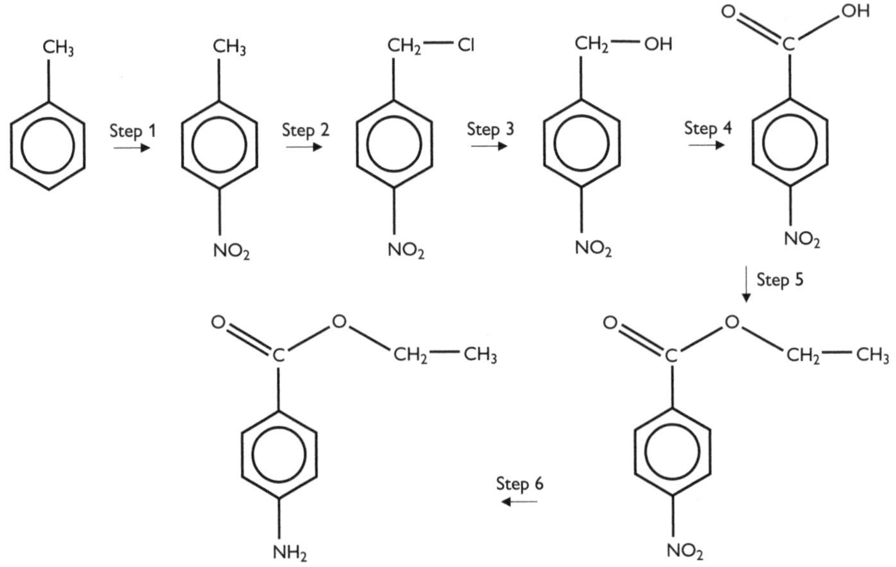

Step 1 → Step 2 → Step 3 → Step 4 → Step 5 → Step 6

Benzocaine

(i) State the types of mechanism for the reactions in steps 1 and 3. (2 marks)

(ii) Classify the reaction types in steps 4 and 5. (2 marks)

(iii) Write balanced equations to describe the reactions in steps 2, 4 and 6. (6 marks)

(iv) State the reagents and conditions used in step 4. (3 marks)

Total: 48 marks

Grade-A answer to Question 12

(a) (i) Step 1 = nitration ✓; step 2 = reduction ✓; step 3 = acylation ✓

The question asks you to name the reaction types rather than the types of mechanisms.

(ii) Step 1 = concentrated nitric acid and concentrated sulfuric acid ✓; step 2 = tin and concentrated hydrochloric acid ✓; step 3 = ethanoyl chloride ✓

You must emphasise that both acids are concentrated in the nitration reaction. In step 2, alternative reagents include iron and concentrated hydrochloric acid or hydrogen with a nickel catalyst. In step 3, ethanoic anhydride could be used.

(iii) Electrophilic substitution ✓

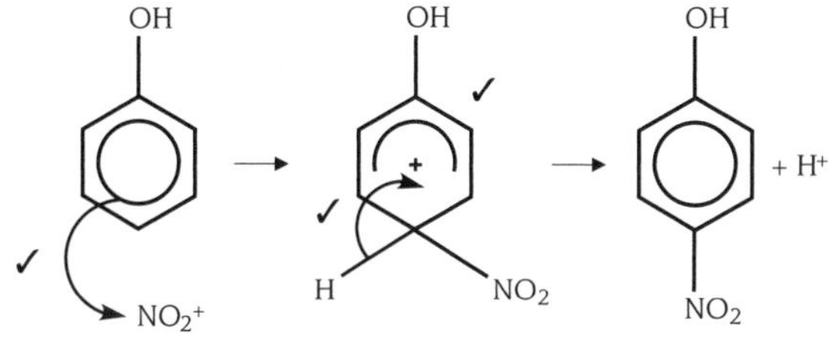

The first mark is for the curly arrow from the ring of delocalised electrons to the N of the nitronium ion (NO_2^+). The second mark is for drawing the structure of the intermediate with the NO_2 group in the 4 position, the broken ring of delocalised electrons and the positive charge in the correct position. The third mark is for the loss of the H^+ with the curly arrow showing the electron pair returning to the benzene to re-form the ring of delocalised electrons.

(b) (i) A = ethanal ✓; B = 2-hydroxypropanenitrile ✓

(ii) Step 1 = oxidation ✓, step 3 = hydrolysis ✓

(iii) Nucleophilic addition ✓

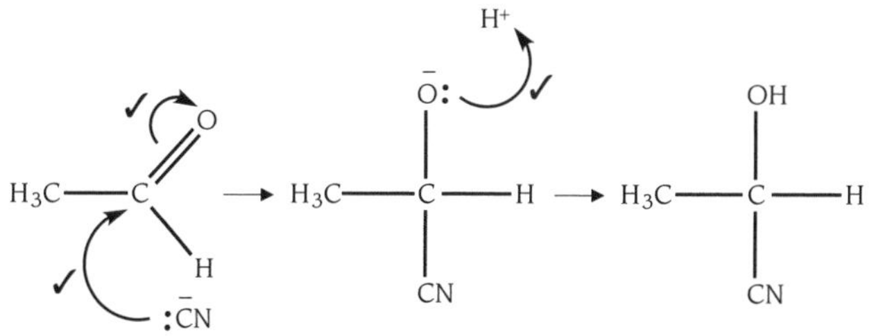

The first mark is for the curly arrow from the cyanide ion to the electron-deficient carbon atom in the carbonyl group. This mark will only be awarded if you have shown the lone pair and the negative charge on the cyanide ion. The second mark is for the curly arrow from the bond in the C=O group to the oxygen atom. The third mark is for drawing the correct intermediate and showing the curly arrow from the oxygen to the H^+ ion. You must show the lone pair and the negative charge on the oxygen atom.

(iv) Potassium dichromate(VI) ✓ and dilute sulfuric acid ✓ under distillation conditions ✓.

It is essential to carry out the reaction under distillation conditions. If reflux conditions were used then the aldehyde would undergo further oxidation to the carboxylic acid.

(v) Step 2 = hydrogen cyanide ✓; step 3 = dilute sulfuric acid ✓

In step 2, HCN is the expected reagent, as stated in the specification. However, an alternative set of reagents would be NaCN and dilute sulfuric acid, which would generate HCN. NaCN on its own is not acceptable. In step 3, dilute hydrochloric acid is acceptable.

(c) (i) Add Tollens' reagent ✓: compound 1 — no reaction ✓; compound 2 — produces a silver mirror ✓

$$H_3C{-}C_6H_4{-}CH_2{-}CHO + [O] \longrightarrow H_3C{-}C_6H_4{-}CH_2{-}COOH$$ ✓

This question is asking you to distinguish between an aldehyde and a ketone, so there are alternative answers. The equation remains the same but the requirements for the first 3 marks are different. Alternative 1: add Fehling's solution and heat — the ketone will give no reaction but the aldehyde will produce a red precipitate. Alternative 2: add acidified potassium dichromate(VI) — the ketone will give no reaction but the aldehyde will change the colour of the solution from orange to green.

(ii) Add bromine water ✓: compound 3 — no reaction ✓; compound 4 — decolorises the bromine ✓

H_3C–(cyclohexene ring)–$C(CH_3)=CH_2$ + $2Br_2$ ⟶ (ring with H_3C, Br, Br)–$C(CH_3)(Br)$–CH_2–Br ✓

This pair of compounds comprises an alkane and an alkene, so the use of bromine water or bromine is the obvious choice. You must show the alkene reacting with 2 moles of bromine in the equation.

(iii) Add sodium hydrogencarbonate ✓: compound 5 — liberates carbon dioxide ✓; compound 6 — no reaction ✓

$$C_6H_5CH_2COOH + NaHCO_3 \longrightarrow C_6H_5CH_2COO^-Na^+ + H_2O + CO_2$$ ✓

The final pair of compounds comprises a carboxylic acid and an ester, so the addition of any metal carbonate or metal hydrogencarbonate to produce carbon dioxide is acceptable. Other alternative answers include the addition of PCl_5, which would produce misty fumes of HCl with the acid but give no reaction with the ester. Addition of a suitable metal, such as magnesium, would produce hydrogen gas with the acid but give no reaction with the ester. In the equation the benzene ring can be represented as in the question or by C_6H_5–.

(d) (i) Step 1 = electrophilic substitution ✓; step 3 = nucleophilic substitution ✓

The question asks you state the type of mechanism and not the type of reaction, so 'nitration' for step 1 and 'hydrolysis' for step 3 would not be accepted.

(ii) Step 4 = oxidation ✓; step 5 = esterification ✓

This question asks for the reaction types and there are no alternative answers.

(iii) Step 2: $CH_3–C_6H_4–NO_2 + Cl_2 \longrightarrow ClCH_2–C_6H_4–NO_2 + HCl$ ✓ ✓
Step 4: $HOCH_2–C_6H_4–NO_2 + 2[O] \longrightarrow HOOC–C_6H_4–NO_2 + H_2O$ ✓ ✓
Step 6: $CH_3CH_2OOC–C_6H_4–NO_2 + 6[H] \longrightarrow CH_3CH_2OOC–C_6H_4–NH_2 + 2H_2O$ ✓ ✓

In each of these equations, the benzene ring can either be represented by a hexagon with a ring of delocalised electrons inside or by $–C_6H_4–$ (not C_6H_6 because two of the hydrogen atoms have been replaced by other substituents). In steps 4 and 6, the oxidising agents and reducing agents are represented by [O] and [H].

(iv) Dilute sulfuric acid ✓ and potassium dichromate(VI) ✓ under reflux ✓

Acidified potassium dichromate(VI) would score the first 2 marks. Reflux conditions ensure that the primary alcohol is completely oxidised to the carboxylic acid.

Question 13

Structure determination

You will need to use the data below in these questions.

Proton NMR chemical shift data

Type of proton	Chemical shift, δ/ppm
RCH_3	0.7–1.2
R_2CH_2	1.2–1.4
R_3CH	1.4–1.6
$RCOCH_3$	2.1–2.6
$ROCH_3$	3.1–3.9
$RCOOCH_3$	3.7–4.1
ROH	0.5–5.0

Infrared absorption data

Type of bond	Wavenumber/cm^{-1}
C–H	2850–3300
C–C	750–1100
C=C	1620–1680
C=O	1680–1750
C–O	1000–1300
O–H (alcohols)	3230–3550
O–H (acids)	2500–3000

(a) An organic compound X has the empirical formula C_2H_4O. It produces the following spectra:

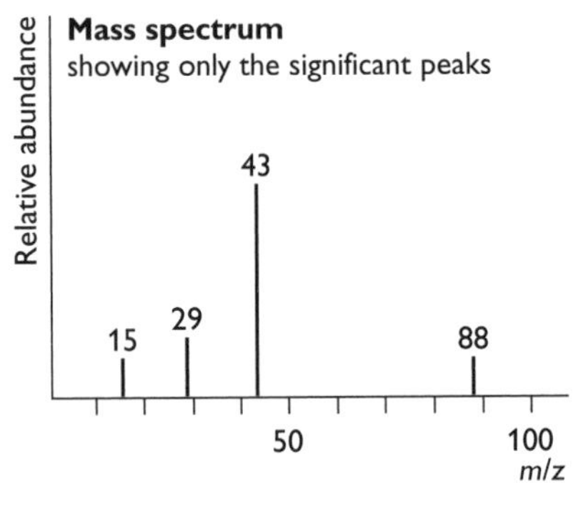

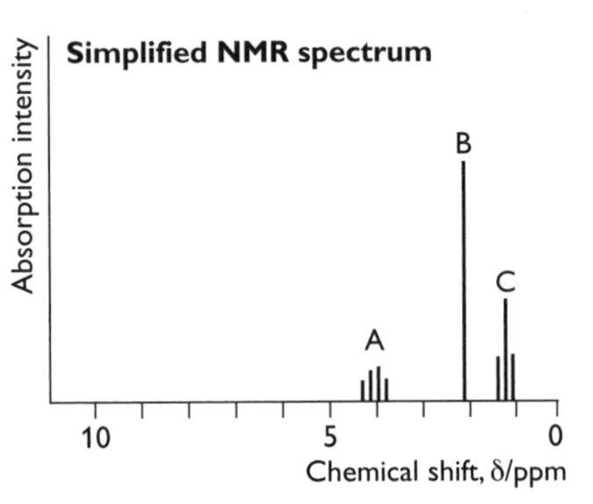

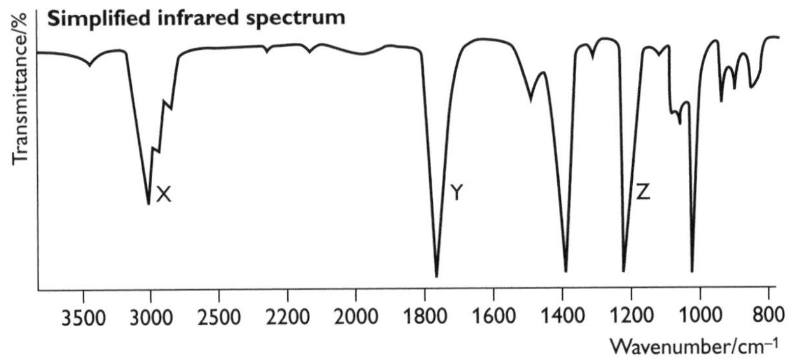

(i) Use the mass spectrum to deduce the molecular formula of compound X. (2 marks)

(ii) Use the table of data to identify the bonds responsible for the absorptions at X, Y and Z in the infrared spectra. (3 marks)

(b) The simplified high-resolution NMR spectrum of compound X shows three signals at A, B and C. The measured integration trace for these peaks gives a ratio of 1.2:1.8:1.8 and the signals occur at chemical shift values of δ = 4.1, δ = 2.1 and δ = 1.2, respectively.

(i) How many different types of proton are present in compound X? (1 mark)

(ii) What is the actual whole number ratio of the numbers of each type of proton? (1 mark)

(iii) Use the table of data to suggest the identity of the group that causes the signal at δ = 2.1. (1 mark)

(iv) The peaks at δ = 4.1 and δ = 1.1 are caused by the presence of an alkyl group. Identify the group and explain the splitting pattern. (3 marks)

(c) Write an equation to show the fragmentation of the molecular ion in the mass spectrometer that leads to the formation of the most abundant peak in the mass spectrum shown in part (a) above. (3 marks)

(d) (i) Draw the structure of compound X. (1 mark)

(ii) Suggest a technique that could be used to separate a mixture of X, b.p. 80°C and an isomer of X, b.p. 77.2°C. (1 mark)

(e) Low-resolution NMR spectroscopy can be used to distinguish between three isomeric carbonyl compounds A, B and C, which gave molecular ions with a mass:charge (*m*/*z*) ratio of 72 in their mass spectra. Compounds A and B both have three peaks in their low-resolution NMR spectra with the ratio of peak areas being 3:2:3 for compound A and 6:1:1 for compound B. In the low-resolution NMR spectrum of compound C there are four peaks, with a ratio of 3:2:2:1. Identify compounds A, B and C. (3 marks)

(f) Compound D is a saturated hydrocarbon, which also gives a molecular ion with an *m*/*z* ratio of 72 in its mass spectrum. The low-resolution NMR spectrum of compound D shows only one peak. Deduce the molecular formula of compound D and draw its structure. (2 marks)

Total: 21 marks

Grade-A answer to Question 13

(a) (i) The largest *m*/*z* value occurs at 88 ✓. This is twice the value of the empirical molecular mass, so the molecular formula must be $C_4H_8O_2$ ✓.

The peak with the largest *m*/*z* value is produced by the molecular ion and gives the relative molecular mass of the compound. The molecular formula must be a whole number multiple of the empirical formula.

(ii) Absorption at X ($3000\,cm^{-1}$) is due to the C–H bond ✓.
Absorption at Y ($1750\,cm^{-1}$) is due to the C=O bond ✓.
Absorption at Z ($1250\,cm^{-1}$) is due to the C–O bond ✓.

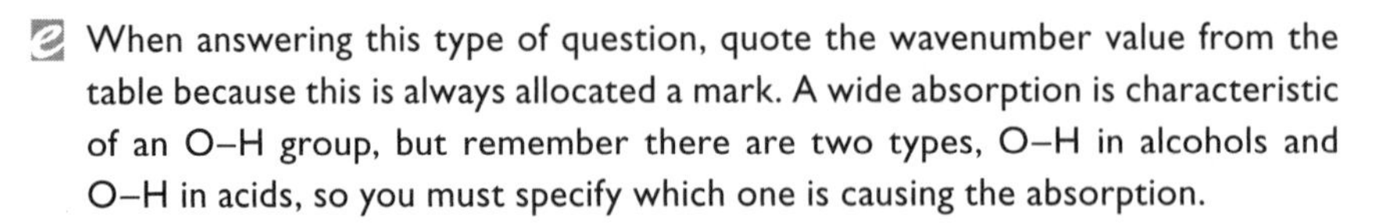

When answering this type of question, quote the wavenumber value from the table because this is always allocated a mark. A wide absorption is characteristic of an O–H group, but remember there are two types, O–H in alcohols and O–H in acids, so you must specify which one is causing the absorption.

(b) (i) There are three different types of proton present ✓.

The number of signals in an NMR spectrum indicates the number of different types or non-equivalent protons present in the compound. In this case, three signals means three different types of proton.

(ii) The ratio is 2:3:3 ✓.

The ratio of 1.2:1.8:1.8 simplifies to 1:1.5:1.5 so the simplest whole number ratio of the protons must be 2:3:3.

(iii) The group causing the signal at $\delta = 2.1$ is RCO**CH_3** ✓.

This is found from the chemical shift values in the table of data. It has a relative intensity of 3 because there are three protons in a methyl group.

(iv) An ethyl group, CH_3CH_2–, causes these signals ✓. The peak at $\delta = 4.1$ is due to the CH_2 protons. It is split into a quartet because there are three equivalent adjacent protons (the CH_3 group) ✓. The peak at $\delta = 1.1$ is due to the CH_3 protons. It is split into a triplet because there are two equivalent adjacent protons (the CH_2 group) ✓.

The peak at $\delta = 4.1$ has a relative intensity of 2 because there are two protons in a CH_2 group. The splitting pattern is predicted using the $n+1$ rule, that is, it has three adjacent protons, so this splits the peak into four (a quartet). The peak at $\delta = 1.1$ has a relative intensity of 3 because there are three protons in a CH_3 group. Again, using the $n+1$ rule, it is split into a triplet by the two adjacent protons in the CH_2 group.

(c) $[C_4H_8O_2]^{+\bullet}$ ✓ ⟶ CH_3CO^+ ✓ + $C_2H_5O^\bullet$ ✓

When writing the equation for the fragmentation of the molecular ion, you must include the 'plus' and the 'dot' to show the radical cation splitting into the cation (which is detected and produces a signal) and the radical (which is not detected).

(d) (i) Ratio 3:2:3 protons from NMR

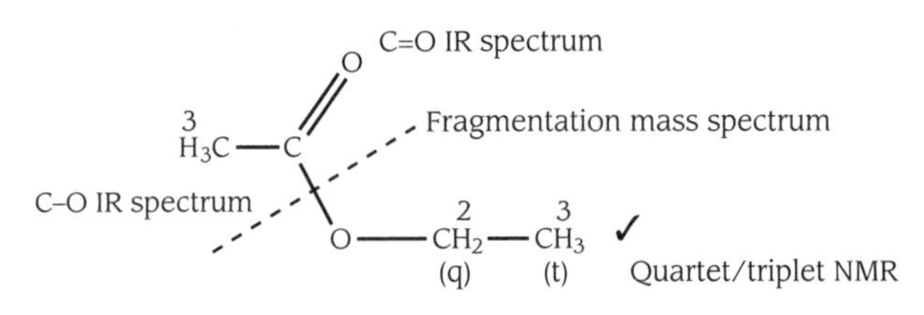

There are two esters that would produce a ratio of 2:3:3 in their low-resolution spectra and a pattern of quartet, singlet and triplet. These are ethyl ethanoate, $CH_3COOCH_2CH_3$, and methyl propanate, $CH_3CH_2COOCH_3$. The structure of ethyl ethanoate is correct because it matches the fragmentation pattern of the mass spectrum. If the compound had been methyl propanoate, it would have fragmented to produce the acylium ion $CH_3CH_2CO^+$ and $CH_3O^•$, and the acylium ion would be detected at an *m*/*z* of 57 rather than at 43, which is caused by CH_3CO^+.

(ii) Column chromatography *or* gas liquid chromatography (GLC) ✓.

Efficient separation of the two liquids could not be achieved by fractional distillation as the boiling points are too close together.

(e) **A** $H_3C-C(=O)-CH_2-CH_3$ ✓ **B** $H_3C-CH(CH_3)-CHO$ ✓ **C** $H_3C-CH_2-CH_2-CHO$ ✓

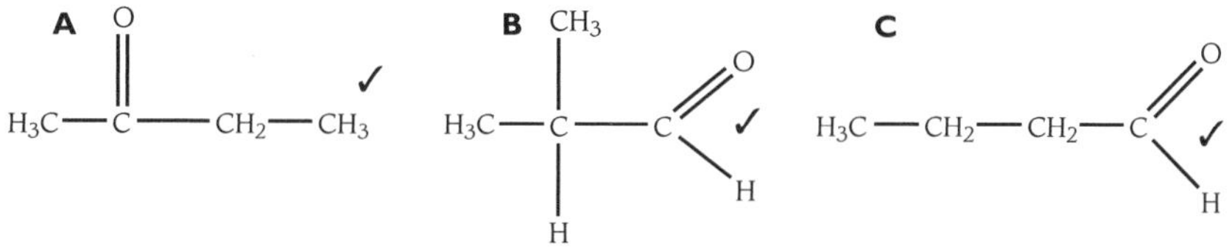

Carbonyl compounds have the general formula $C_nH_{2n}O$. A molecular mass of 72 indicates a molecular formula of C_4H_8O, that is $(4 \times 12) + (8 \times 1) + (1 \times 16) = 72$. There is only one ketone with the molecular formula C_4H_8O, which is butanone. Draw the aldehyde with the longest carbon chain to identify butanal; branching identifies the other aldehyde as methylpropanal. Once the compounds have been identified, they must be matched to the information provided by the low-resolution NMR spectra.

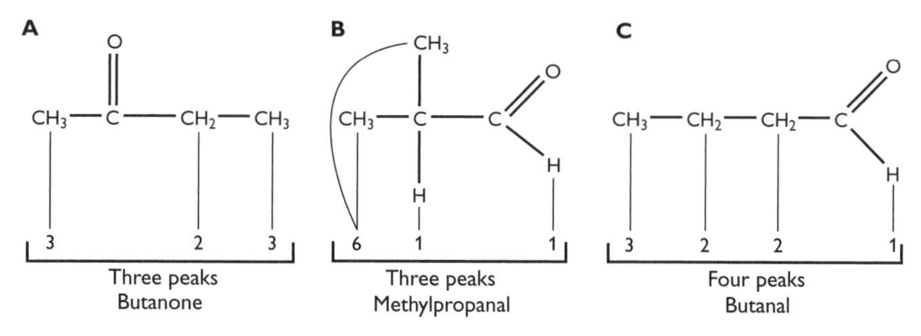

(f) The molecular formula is C_5H_{12} ✓.

The structure of D is: $C(CH_3)_4$ ✓

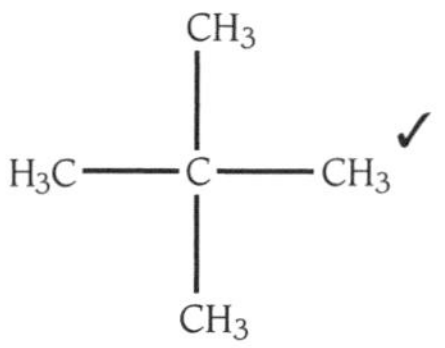

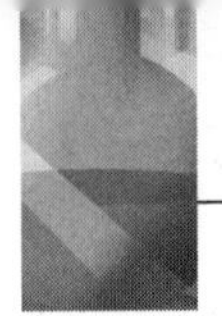

The general formula of a non-cyclic saturated hydrocarbon is C_nH_{2n+2}. Since the M_r is 72, the molecular formula is C_5H_{12}, that is (5 × 12) + (12 × 1) = 72. The compound must have four methyl groups because all the hydrogen atoms must be in the same environment.